Springer
Berlin
Heidelberg
New York
Barcelona
Budapest
Hongkong
London
Mailand
Paris
Santa Clara
Singapur
Tokio

Francis P. Gasparro (Ed.)

Sunscreen Photobiology:

Molecular, Cellular and Physiological Aspects

Springer

Francis P. Gasparro, Ph.D.
Thomas Jefferson University
Department of Dermatology
and Cutaneous Biology
233 South 10th St. (BLSB 428)
Philadelphia, PA 19107
U.S.A.

ISBN 3-540-63251-4 Springer-Verlag Berlin Heidelberg New York
Biotechnology Intelligence Unit

Library of Congress Cataloging-in-Publication Data

Sunscreen photobiology: molecular, cellular, and physiological aspects / [edited by] Francis P. Gasparro.
p.cm. - (Biotechnology intelligence unit)
Includes bibliographical references and index.
ISBN 1-57059-468-6 and 3-540-63251-4 (alk. paper)
1. Sunscreen (Cosmetics). 2. Sunscreens (Cosmetics)-Toxicology. I. Gasparro, Francis P. II. Series
RS431.S94S866 1997
612'.01444-dc21 97-21826 CIP

Typesetting: Landes Bioscience Georgetown, TX, U.S.A.

SPIN:10634495 31/3111 - 5 4 3 2 1 0 - Printed on acid-free paper

Dedication

To all my teachers who taught me to think critically and challenge conventional wisdom. Among these is my father, who passed away as this book neared its completion. Thank you.

PREFACE

This monograph is designed to provide a snapshot of some critical issues in the field of sunscreen photobiology. While it may be common to assume that sunscreen molecules are photochemically inert and unreactive, it cannot be forgotten that any light-absorbing molecule has the potential for reactivity by simply appreciating the first law of photochemistry (first stated in 1818 by Grotthus and Drapper). In another perspective it must be appreciated that no sunscreen, even if photochemically inert, can absorb all wavelengths of light that might alter underlying skin biochemistry. Thus there can be indirect sunscreen photobiology as a result of light transmitted by sunscreen protected skin (SPS).

This book is not meant to be a comprehensive treatise on skin. Excellent descriptions of skin physiology and biochemistry may be found in many dermatology texts. The aim here is to present cutting-edge issues. Hence there are reviews of action spectra (chapter 1) and sunscreen photophysical and photochemical properties (chapters 2 and 3). Chapter 4 surveys normal effects of sunlight on skin. Chapters 5 and 6 describe novel aspects of photoaging and highlight how for the axiom—there's nothing new under the sun—is often untrue! Chapters 7 and 8 deal with the effects of sunlight on the immune activity of the skin. Each set of authors make potent and cogent arguments for their viewpoints. It is expected that the readers of this volume will come to informed conclusions of their own. A more complex picture and food for additional thought is presented in chapter 9.

A personal note about chapter 10 is in order here. I have been fascinated by the topic of this book ever since I first irradiated a buffered PABA solution with UVB in 1978. Burned into my mind is the image of the brick red solution that resulted. The photochemistry of PABA now includes its photoreactions with thymine (Martin Shetlar and co-workers; see *Photochem Photobiol* 1992; 55: 657- 663). The argument that sunscreens don't get into cells just does not hold water. When sunscreens are used at the recommended levels of 2 mg/cm^2, there is a tremendous excess of sunscreen molecules to cells (by several orders of magnitude). Even if only a minute fraction of these get into cells (for an example see M-H Tan et al, *Australas J Dermatol* 1996; 37: 185-187), it is feasible that a few sunscreen-mediated events (photoadduct formation or reactive oxygen species generation) could lead to deleterious effects.

Over these nearly two decades the complexity of skin biology has become more fully appreciated. In addition, we have witnessed the impact of developments in molecular biology on our understanding of events in skin. The transition from skin physiology and biochemistry to molecular biology means that events mediating processes ranging from cytokine signaling to skin pigmentation and cancer induction can be traced to distinct molecular steps. Finally, at this time we are poised on the footsteps of a new frontier. Genetic therapy should revolutionize the treatment of many human diseases. Given the accessibility of the skin, it would not be surprising to find the first breakthroughs in the treatment of skin diseases.

The first chapter by R. Setlow (Brookhaven National Laboratory) summarizes our understanding of the wavelengths of light that affect skin biology and thus defines those wavelengths that a sunscreen must absorb in order to protect skin. The erythema action spectrum is the best understood and hence, designing sunscreens to prevent sunburn is the most knowledge-based. But we need to know more; specifically, which wavelengths of light (radiation) are responsible for skin immune suppression and skin cancer. Regarding the latter, we need to know what is causing the *epidemic* of skin cancer. We must remember that no sunscreen completely blocks all wavelengths of light (including so-called *physical* sunblocks). We also need to understand the skin photobiology that results from the spectrum of sunlight transmitted through sunscreen protected skin (SPS). The SPS spectrum will depend strongly on the sunscreen ingredients, but it is certain that it will contain longer UVA wavelengths as well as most of the visible spectrum of sunlight.

John and Sandra Allen (Indiana State University) and Bice Martincigh (the University of Natal) describe the physical and chemical properties of the current group of FDA-approved compounds from which selections are made to formulate sunscreen products. If one starts from the premise that sunscreens are not photoreactive, the challenge here regards the effect of the SPS spectrum—namely what wavelengths of light must be absorbed by an ideal sunscreen product to not only prevent sunburn but also to prevent immune suppression or any other deleterious effect. The authors also describe the ability of sunscreens to generate singlet oxygen and to sensitize the formation of thymine dimers.

In chapter 3 John Knowland (University of Oxford) discusses the potential for photoactivated sunscreens to damage DNA. Such events have been demonstrated *in vitro* by several investigators. What remains to be determined is whether some of these reactions also occur *in vivo*. But now this becomes a 'needle in the haystack' problem. Deleterious effects could be initiated by a single hit in a specific gene. In fact it could be argued that sublethal damage in SPS could exacerbate this effect. The now near-classical example of this has been repeatedly demonstrated in the fingerprints of particular agents in the mutation spectrum of p53 from a wide range of cancers, including skin cancer. Regarding the latter, this story has been recounted many times—a particularly good summary by D. Brash and D. Leffell may be found in the August 1996 issue of *Scientific American*.

An important issue raised in both chapters 2 and 3 concerns the properties of so-called physical sunscreen titanium dioxide. This classical pigment (a chemical!) in its currently used form is very different from the opaque form we all visualize on the noses and lips of lifeguards. In order to make TiO_2 cosmetically acceptable for widespread use by the public, it has been micronized. The reduced particle size changes its physical (and perhaps chemical and physiological) properties. Titanium dioxide is a semiconductor and as such can initiate photochemical processes after the absorption of a wide range of wavelengths of light, including longer UVA wavelengths and the shorter wavelengths of visible light.

The next three chapters describe the effects of light on skin. In chapter 4 Cheryl Rosen (University of Toronto) summarizes the known visible effects of light on unprotected human skin (erythema and tanning). In chapter 5 Lorraine Kligman (University of Pennsylvania) describes the effects of chronic irradiation on skin protected with three different strength sunscreens (SPFs of 7, 14 and 18). Her work dramatizes the effects of light penetrating SPS. The higher SPF sunscreens offered the greatest protection as gauged by gross analysis of tissue. These studies clearly illustrate the effects of SPS transmitted light on skin. These studies also illustrate another point. Delineating light-induced effects in animals requires chronic exposures and can take a good part of a year to complete. A potential solution to time-consuming animal studies is described in chapter 6. Eric Bernstein (Thomas Jefferson University) developed a transgenic mouse with a human elastin promoter-CAT construct. This system has been used to exhibit dose-dependent induction of the elastin promoter (measured by chloramphenicol transferase activity) after single doses of either UVB or 8-MOP plus UVA. Relevant to the topic of this monograph, the system has been used to demonstrate the protective effects of sunscreens as indicated by a reduction in the level of CAT activity in SPS. The molecular readout provides a major advantage over the usual physiological readout in conventional animal studies because it can be obtained after a single phototreatment (*in vitro* or *in vivo*). Although the model was developed to evaluate events involved in photoaging, the resultant elastin-CAT activity can serve as a surrogate molecular marker for photochemical events in skin. There is no doubt that other molecular markers may be similarly induced; these are discussed in chapter 10.

At this point with current sunscreens there is no question about the ability of sunscreens to protect skin from burning. They are selected and formulated with this as their primary function. When pioneering work in the early 1980s revealed that the skin immune system could be compromised by sunlight, it was necessary to show that sunscreens also prevented immune suppression. Many studies have since been published.

Two viewpoints and up-to-date summaries of the effect of sunscreens on UV-irradiated skin immune properties are given in chapters 7 and 8. In chapter 7 Peter Wolf (University of Graz) and Margaret Kripke (MD Anderson Cancer Center) succinctly summarize the fundamentals of skin immunology. They also survey potential mechanisms of UV-induced immune suppression and the corresponding action spectrum. They note that most of the studies they cite have been performed in animals and point out the need to ultimately obtain similar data in humans. The limited literature regarding the latter is also summarized. Their chapter concludes with two very important sections. First, they describe seven reasons why sunscreens appear to provide insufficient immune protection. Second, they discuss future aspects of suppression as related to sunscreens. A central motif is an appreciation of the impact of photochemistry of different cellular entities on skin biology.

In chapter 8 Lee Roberts and Douglas Learn provide a critical analysis of the need for carefully designed experiments. Primary among their concerns is that the source of light used in immune studies must match the power spectrum of sunlight. In addition, they emphasize the importance of recognizing differences between human skin (where sunscreens are used) and mouse skin (where for ethical reasons sunscreens are developed and tested). These two chapters provide a point-counterpoint perspective on this important aspect of sunscreen photobiology.

Vivienne Reeve (University of Sydney, Australia) introduces the concept of urocanic acid and its potential relationship with sunscreen protection, immune suppression and photocarcinogenesis in mice. Reeve points out that the incidence of both nonmelanoma skin cancer and malignant melanoma have doubled between 1985 and 1995 in Queensland. While the effects of increased sunscreen use would not be expected to attenuate the cancer initiation process, it is not unlikely for a truly effective sunscreen to prevent the promotional steps. A new term is presented—SSUV—sunscreened ultraviolet radiation, a UVA-rich spectrum of light that penetrates sunscreen protected skin. Furthermore, the surprising efficacy of padimate-O is described and it is suggested that it or a new derivative be reinvestigated for use in sunscreens for human skin protection. This chapter also introduces the field of urocanic acid isomerization and its role in skin immune suppression. Finally, it is emphasized that the measurement of immune function by an acute test may not apply in chronically irradiated skin.

While the protective effects of sunscreens are clearly well established, there can be no doubt that the spectrum of light penetrating SPS may have biological effects. The evidence comes from observations of biological effects induced in skin cells by suberythemogenic doses of UV light. In chapter 10 three examples are presented—p53, NFκB and AP-1. The first was chosen because of its great impact on the fate of cells. Properly functioning p53 directs a damaged cell down an apoptotic pathway and renders it unable to affect the skin in an untoward way. The latter are mentioned as additional paradigmatic examples of what might be viewed as a counterintuitive kind of effect—the turning on of genes by cell damage. Other kinds of molecular marker, in cells are the mutations that result from UV exposure. Ultimately these may lead to cellular transformations and the development of skin cancer. While an attractive added feature of sunscreens would include the ability to protect against skin cancer, it remains to be an established fact (in humans). In order to address the apparent confounding connection between increased incidence of skin cancer and the greater use of sunscreens, the question of potential endogenous long wavelength UVA chromophores are discussed. Additionally, the not too extensive literature describing iron deposition in chronically irradiated skin is briefly summarized.

As I stated in the opening section of this Preface, this book is a snapshot of today's critical issues in sunscreen photobiology. As the authors of these chapters continue their endeavors and as others are challenged by these exciting issues, there will be inevitable progress in our understanding of sunscreen photobiology. In a few years a new snapshot will no doubt portray a different set of critical issues.

Francis P. Gasparro
Philadelphia, PA
April 14, 1997

CONTENTS

1. **Action Spectra: Skin** 1
Richard B. Setlow
Introduction 1
The Determination of Action Spectra 1
Examples of Problems in Interpreting Action Spectra 4
Conclusion 8

2. **Sunscreens: The Molecules and Their Photochemistry** 11
Bice S. Martincigh, John M. Allen and Sandra K. Allen
Introduction 11
Photochemical Behavior of Chemical Absorbers 28
Photochemical Formation of Reactive Species by Physical Blockers 40
Additional Considerations 41
Conclusions 42

3. **The Photochemical Potential of Some Sunscreens to Damage DNA** 47
John Knowland, Peter J. McHugh and Rosemary Dunford
Introduction 47
Background 48
Chemical Sunscreens 50
'Physical' Sunscreens 55
Conclusions 59

4. **Sunscreens and the Prevention of Erythema** 63
Cheryl F. Rosen

5. **Do Broad Spectrum Sunscreens Allow Ultraviolet-Induced Photodamage in the Absence of Erythema?** 69
Lorraine H. Kligman
Introduction 69
What Are The Issues? 70
The Rationale for This Study 71
Materials and Methods 72
Results 73
Discussion 79

6. **A Transgenic Mouse Model of Cutaneous Photoaging Measuring Elastin Promoter Activation** 85
Eric F. Bernstein
Intrinsic Versus Extrinsic Aging 85
Extracellular Matrix 86
Photoaging 88
Animal Model of Cutaneous Photoaging Measuring Elastin Promoter Induction 89

7. **Immune Aspects of Sunscreens** 99
Peter Wolf and Margaret L. Kripke
Introduction 99
UV-Induced Immune Suppression and UV Carcinogenesis 100
UV Radiation and Infectious Diseases 102
Mechanisms of UV-Induced Immune Suppression 104
Action Spectrum of UV-Induced Immune Suppression 105
Composition and Concentration of Sunscreens Used for Immune Protection 106
Effect of Sunscreens on UV-Induced Immunologic Alterations in Laboratory Animals 106
Effect of Sunscreens on UV-Induced Immunologic Alterations in Humans 113
Reasons for Insufficient Immune Protection by Sunscreens 116
Future Aspects of Sunscreens and Immune Protection 118

8. **Sunscreen SPF Values and Immune Protection Levels Are Equivalent When Tests Are Conducted by Appropriate Methods and Procedures** 127
Lee K. Roberts and Douglas B. Learn
Introduction 127
The Basics of Sunscreen Protection 128
Review of Sunscreen Immune Protection Studies 132
Conclusions 148

9. **The Relationship Between Sunscreen Protection from Erythema, Photoimmunosuppression and Photocarcinogenesis in the Hairless Mouse** 157
Vivienne E. Reeve
Introduction 157
Commercial Sunscreens and UVB Radiation 159
Experimental SPF 6 Sunscreens and Simulated Solar UV (SSUV) Radiation 165
Summary 170
Conclusions 172

10. **Epilogue: New Perspectives in Sunscreen Photobiology** 177
Francis P. Gasparro
Rationale for This Chapter 177
Molecular Markers 178
Induction 179
Epigenetic Effects 181
Iron Deposition in Skin Chronically Exposed to UV 182
Summary 183

Color Insert 187

Index 191

EDITOR

Francis P. Gasparro, Ph.D.
Photobiology Laboratory
Department of Dermatology and Cutaneous Biology
Thomas Jefferson University
Philadelphia, Pennsylvania, U.S.A.
chapter 10

CONTRIBUTORS

John M. Allen, Ph.D.
Department of Chemistry
Indiana State University
Terre Haute, Indiana, U.S.A.
chapter 2

Sandra K. Allen, Ph.D.
Department of Health and Safety
Indiana State University
Terre Haute, Indiana, U.S.A.
chapter 2

Eric F. Bernstein, M.D.
Department of Dermatology and Cutaneous Biology
Thomas Jefferson Medical College
Philadelphia, Pennsylvania, U.S.A.
chapter 6

Rosemary Dunford, Ph.D.
Department of Biochemistry
University of Oxford
Oxford, U.K.
chapter 3

Lorraine H. Kligman, Ph.D.
Research Professor
Department of Dermatology
University of Pennsylvania
School of Medicine
Philadelphia, Pennsylvania, U.S.A.
chapter 5

John Knowland, Ph.D.
Department of Biochemistry
University of Oxford
Oxford, U.K.
chapter 3

Margaret L. Kripke, Ph.D.
Department of Immunology
The University of Texas
M.D. Anderson Cancer Center
Houston, Texas, U.S.A.
chapter 7

Douglas B. Learn, Ph.D.
Senior Scientist
Research and Development
Schering-Plough HealthCare Products, Inc.
Memphis, Tennessee, U.S.A.
chapter 8

Peter J. McHugh, Ph.D.
Department of Biochemistry
University of Oxford
Oxford, U.K.
chapter 3

Bice S. Martincigh, Ph.D.
Department of Chemistry and Applied Chemistry
University of Natal
Durban, South Africa
chapter 2

Vivienne E. Reeve, Ph.D.
Department of Veterinary Pathology
University of Sydney
N.S.W., Australia
chapter 9

Lee K. Roberts, Ph.D.
Research Director
Research and Development
Schering-Plough HealthCare Products, Inc.
Memphis, Tennessee, U.S.A.
chapter 8

Cheryl F. Rosen, M.D., FRCPC
Women's College Hospital
University of Toronto
Toronto, Canada
chapter 4

Richard B. Setlow, Ph.D.
Biology Department
Brookhaven National Laboratory
Upton, New York, U.S.A.
chapter 1

Peter Wolf, M.D.
Department of Dermatology
University of Graz
Graz, Austria
chapter 7

ABBREVIATIONS

AIDS	acquired immunodeficiency syndrome
BCG	Bacillus Calmette Guérin
CHS, CH	contact hypersensitivity
DNCB	dinitrochlorobenzene
DNFB	2,4-dinitrofluorobenzene
DTH	delayed type hypersensitivity
FDA	U.S. Food and Drug Administration
IL	interleukin
HIV	human immunodeficiency virus
HSV-1	herpes simplex virus type 1
HSV-2	herpes simplex virus type 2
IPF	immune protection factor
I/S	immunosuppression
ISD	immunosuppression dose
LC	Langerhans cells
MECLR	mixed epidermal cell-lymphocyte reaction
MED	minimal erythema dose
MEdD	minimum edema dose
MLR	mixed lymphocyte reaction
o-PABA	octyl-N-dimethyl-p-aminobenzoate
SPF	sun protection factor
SSUV	solar simulated UV
TAA	tumor-associated-antigens
TGF-β1	transforming growth factor-β1
TISD	tumor-immune suppression dose
TNCB	trinitrochlorobenzene
TNF-α	tumor necrosis factor-α
Ts-Cells	suppressor T-lymphocytes
UCA	urocanic acid
UV	ultraviolet radiation
2-EHMC	2-ethylhexyl-p-methoxycinnamate

CHAPTER 1

Action Spectra: Skin

Richard B. Setlow

Introduction

The quantitative determinations of the effects of different wavelengths or spectral regions in sunlight on skin and on molecules in skin are important not only to devise appropriate protective measures against the deleterious actions of light exposure, but also to assess the molecular bases of the effects mediated by molecular absorbers. Such absorbers include the macromolecular structures of nucleic acids, proteins, and lipids, as well as smaller, possibly key, molecules such as urocanic acid and numerous photodynamic sensitizers as well as psoralens.

An action spectrum for affecting skin represents the sensitivity of a response as a function of incident radiation fluence, in energy per unit area. In principle, one should use equal fluence rates at all wavelengths. In practice, except at extremely low or high fluence rates, the response at a given fluence is not dependent on rate and will be ignored for this discussion. Earlier comprehensive reviews of action spectra have been written by Setlow[1] and by Coohill.[2] Action spectra may be divided into phenomenological and analytical ones. The former represents the response of a tissue, for example erythema or cancer induction, although the particular light absorbing component may not be known. If the action spectra for affecting molecules are known, and for simple ones may be taken as similar to the absorption spectra, one attempts to interpret the phenomenological spectra into molecular components—analytical spectra.

The Determination of Action Spectra

To construct an action spectrum one chooses a biological endpoint (Minimal Erythema Dose [MED] at 24 hours, 25% of animals

Sunscreen Photobiology: Molecular, Cellular and Physiological Aspects,
edited by Francis P. Gasparro.

with tumors after daily exposures for 1 year, inactivation of 90% of bacteria, etc.) and determines the fluence at a number of different wavelengths or wavebands so as to achieve the desired endpoints. The reciprocals of the fluence for equal responses are the measures of sensitivity—a low fluence meaning a high sensitivity. A plot of the sensitivity—the reciprocal of the fluence to achieve a particular endpoint—is the action spectrum. To compare one spectrum with another, action spectra are often normalized to the value 1.00 at the wavelengths of maximum sensitivity, or to 1.00 at a common wavelength, e.g., 300 nm. In the latter case, some sensitivities may be greater than 1.00. An action spectrum should *not* be constructed from the responses at equal fluence for a number of wavelengths, because often the response may be a complicated, non linear function of fluence.

Attempts to interpret phenomenological spectra in terms of molecular components are doomed to failure unless the location(s) of the molecules in or below skin is known and a correction made to the incident fluence for the transmission of skin. The complications are that the transmissions may be very low and, especially in the UVB range, are a strong function of wavelength. Hence, corrections to the incident fluence (see below) are so large and variable that they may introduce a big error into the calculated spectral shape. Three cases for the evaluation of action spectra in terms of molecular absorbers are of interest. The first is an optically thin molecular layer below a fixed absorbing layer. The second is the distribution of molecules throughout a stirred absorbing solution, and the third is the distribution of molecules within an absorbing but unstirred medium, such as skin. The first complication is easily corrected for if the location of the absorbers are known and the transmission of the overlying (skin) layer is known as a function of wavelength. One multiplies the incident fluence by the transmission to obtain the corrected fluence to use in constructing an action spectrum.

In the second, stirred, case the molecules occupy random positions and are exposed to the average fluence through the depth of the solution. If the transmission, I/I_0, of monochromatic light through the solution is represented by $e^{-\alpha \underline{l}}$, where α, the absorption coefficient per unit length, is a function of wavelength and $\underline{l}$ is the depth of the solution, the average intensity $\bar{I}$ through the solution is easily shown to be[3]

$$\bar{I} = I_o \frac{(1 - e^{-\alpha \underline{l}})}{\alpha l} = kI_o$$

Table 1.1. Correction factors, k, to convert incident to the average fluence through an absorbing, stirred solution. [3]

Transmission ($T = I/I_o$)	αl ($2.30 \log_{10} I_o/I$)	Correction factor (k)
1.00	0.000	1.00
0.80	0.223	0.90
0.50	0.69	0.72
10^{-1}	2.30	0.39
10^{-2}	4.60	0.22
10^{-4}	9.21	0.11

Hence, the average intensity or fluence through the sample equals the incident intensity or fluence multiplied by a correction factor k equaling $(1 - ea^{l})/al$. This equation holds regardless of the shape of the dose response curve. Table 1.1 shows values of the correction factor *k* for different relative transmissions.

The unstirred absorbing cases are much more complicated because molecules on the surface receive a larger fluence than those below the surface. All the molecules are not exposed to the same dose as they are in the stirred case. As a result, the final fluence-response curve represents a mixture of molecules exposed to different fluence and the overall response curve will be intermediate between that of the most highly and the least highly exposed portions of the sample. For transmissions greater than ~ 20% the correction factor to the incident fluence is slightly less than that for a stirred sample provided that the fraction of affected molecules is no greater than 65%. The difficulties of using action spectra to correlate biological effects with molecular receptors are a function of wavelength for the reasons given above. An appreciation of the penetration depth in skin as a function of wavelength (Table 1.2) is obviously necessary.

If the shape of an action spectrum changes with the response level chosen for the biological endpoint, the action spectrum is not a unique shape. It may represent two or more different processes whose relative magnitudes change with wavelength. For example, in mice chronically exposed to low UV exposures, the ratio of induced papillomas to induced sarcomas is greater in the UVA region compared to the UVB.[5] This result may be taken to indicate that

Table 1.2. The decrease in the incident energy density in skin with depth, in mm, for several wavelengths. [4]

Wavelength (nm)					
250	280	300	350	400	Energy density
Depth (mm)					
0	0	0	0	0	1.00
2	1.5	6	60	90	0.37
4.5	3.5	14	140	210	0.10
9	7	28	260	420	0.01

photosensitized effects in the UVA are of increased importance compared to the effects attributable to the direct absorption of light by DNA.

Examples of Problems in Interpreting Action Spectra

The difficulties in interpreting action spectra obtained for effects in highly absorbing media—solutions or skin—are exemplified by three examples: (1) the isomerization of the trans to cis forms of urocanic acid (UCA) molecules in skin, implicated in immunosuppression; (2) the theoretical construction of an action spectrum for skin cancer induction under the assumption that light absorbed directly in DNA is the initiating act; and (3) the extrapolation of action spectra from animals to humans in the UVA region, where the direct absorption of light by DNA is negligible.

Isomerization of UCA

An action spectrum for the immunosuppression of contact sensitivity in mouse skin has a shape similar to the absorption spectrum of UCA.[6] In an effort to tie down the association of immunosuppression with isomerization of UCA, Riggs et al[7,8] measured the action spectrum for isomerization in vitro and in vivo using the incident fluence as the experimental parameter. The technique for the in vitro spectrum was to irradiate solutions of UCA at a 6 mM concentration of the trans-isomer—an approximation to that in skin. This was a reasonable approach to standardizing the in vitro and in vivo reactions. However, in vivo the UCA is in a skin thickness less than ~ 0.1 mm while in vitro it was in a stirred solution of 10 mm thickness. The authors appreciated the necessity of stirring the absorbing solutions, but in their calculations of the action spectrum they used the incident fluence rather than the average fluence through the stirred sample. At 270 nm UCA has a molar extinction coefficient of approximately 16,000 $M^{-1}cm^{-1}$ (ref. 9) and hence a 1 cm, 6 mM solution would have an absorbency of approximately 100, or a transmittance of 10^{-100}, whereas at 340 nm the transmittance is approximately 0.9. Since 10^{-100} equals e^{-230}, the appropriate correction factor to the incident energy (Equation 1.1) is 1/230. The correction factor at 340 nm is 0.95. Hence, the action spectrum value at 270 nm is too small by 230-fold. I used the UCA absorption spectrum,[9] the data in Riggs et al[7] and Equation 1 to obtain the corrections to the action spectra data points and from them the corrected action spectrum. The latter is shown in Figure 1.1. The corrected action spectrum points, after making an adjustment for the fact that at given energy fluence there are fewer photons at short wavelengths than at low ones—a so-called quantum correction—were normalized to 1.00 at 270 nm. The magnitude of the difference between the corrected and the incorrect spectrum indicates clearly the need to measure, and correct for, the absorbency of the irradiated solution. The agreement between the corrected in vitro action spectrum and the absorption spectrum of UCA is striking, however, three caveats are important: (1) The use of an absorption spectrum, computed from several concentrations, as a source of data to make corrections to the incident energy does not give as reliable result as direct measurement of the transmission at different wavelengths; (2) The similarity between the action spectrum and the absorption spectrum implies that the quantum yield for isomerization is independent of wavelength. However, direct measurements indicate that this is not the case.[10] It increases by a factor of approximately 10 between 264 and 313 nm. The discrepancy between these two ways of estimating quantum yields may be the result of using an improper absorbency of a 6 mM UCA solution at a shorter wavelength; (3) The action spectrum in vivo cannot be readily interpreted unless one can make a correction for the decrease in light transmission as a function of depth (case 3 above).

Cellular DNA as a Target for Carcinogenesis by Light

There is good evidence indicating that DNA is the most important target molecule for the initiation of cancer. For example,

Fig. 1.1. The action spectra (solid points) and the absorption spectrum redrawn from figure 1 of ref. 7, and the corrected in vitro action spectrum (open points) for UCA isomerization (R. Setlow, unpublished). The correction was made using published absorption spectra[9] and the reported concentration, 6 mM, of the irradiated UCA solution.[7]

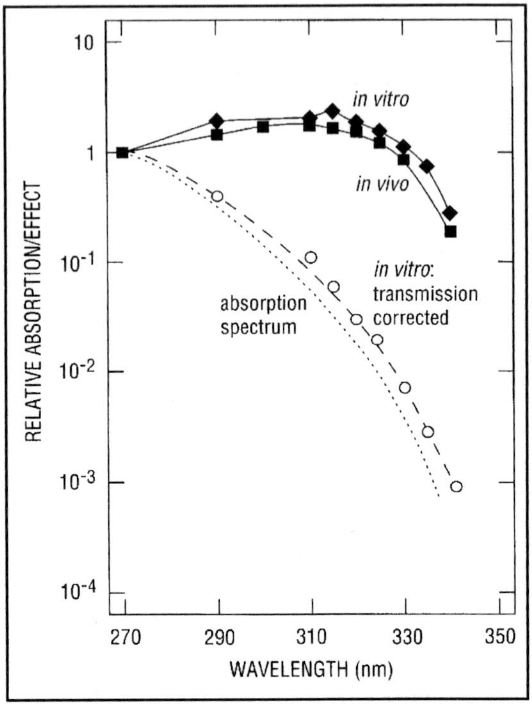

individuals with the genetic disease xeroderma pigmentosum are orders of magnitude more susceptible to sunlight-induced skin cancer and they are deficient in repairing UV-damaged DNA by a process called nucleotide excision repair.[11,12] Hence, one infers that the action spectrum for skin cancer induction should show a very large component representing the action spectrum for affecting DNA directly. Such a spectrum was compiled from an average of many spectra representing photoproduct formation, the inactivation of DNA viruses, and bacterial killing and mutagenesis.[13] One assumes that such a spectrum would hold for the DNA in skin or in plant cells. Unfortunately, this spectrum, named after this author, has been used by others in many publications without noting, as the original publication did, that DNA is not on the surface of skin or of plants. The spectrum must be corrected for the transmission[4] of the overlying tissue by multiplying the spectrum by the transmission at the wavelength used. This corrected action spectrum is similar to the spectrum for forming cyclobutane dimers in human skin, irradiated in vivo when one averages out the large interindividual variations that presumably arise from differences in transmissions among

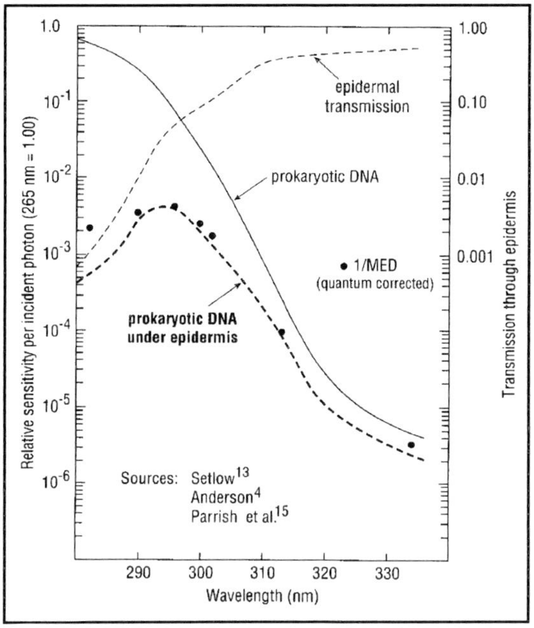

Fig. 1.2. Action spectra for affecting prokaryotic DNA,[13] prokaryotic DNA under epidermis corrected for epidermal transmission[4] (R. Setlow, unpublished) and human minimal erythema dose (MED),[15] solid circles.

individuals.[14] The transmission is negligible at wavelengths in the neighborhood of 260 nm. As a result the action spectrum for damaging DNA in skin has a maximum at approximately 300 nm rather than at 260 nm as observed for pure DNA. Figure 1.2 shows the uncorrected and corrected spectra along with the action spectrum for erythema induction—a phenomenological spectrum needing no correction.[15] The DNA spectra are expressed as the reciprocal of the photons per unit area normalized to 1.00 at 260 nm. The erythemal spectrum is in m^2/J, with a slight adjustment to allow for the fact that there are more quanta per Joule at the longer wavelengths than at 300 nm, taken as the uncorrected wavelength.

If one observed a carcinogenic action spectrum, at wavelengths greater than 300 nm, similar to the corrected DNA spectrum, one would be hard pressed to interpret the action spectrum as indicating an important role for DNA in carcinogenesis. Although the spectrum is consistent with DNA as a target, one would have to go through the reverse process of correcting the observed spectrum by dividing it by the transmission of skin as a function of wavelength to obtain the molecular spectrum. By the same token, the similarity between the observed erythemal spectrum and the corrected DNA spectrum cannot be taken to prove that DNA damage is the molecular target for erythema without other data—data such as obtained in the marsupial, *Monodelphis domestica*, indicating that erythema is photoreactivable,[16] a reaction that reverses UV-induced photoproducts in the skin DNA of this animal.

UVA Effects

At long UV wavelengths, $\lambda > 330$ nm, the action spectra for human erythema,[15] squamous cell carcinoma induction in mice,[17] melanoma induction in fish,[18] and mutagenicity in human cells in culture[18] do not fall off as rapidly as one would predict for the direct effect on DNA. The presence of small absorbing molecules that transfer energy to DNA, or that participate in oxidizing reactions that may affect DNA, is inferred. To date action spectra have not identified these presumed molecular targets involved in the indirect effects on DNA. The best that has been done in light-induced cancer is to extrapolate the mouse skin cancer data to humans by correcting for the low transmission of human skin compared to mouse.[19] This correction should certainly work for the UVB region. However, its use in the UVA, although appropriate, involves the added assumption that the sensitizing chromophores in humans are quantitatively similar to those in mice. A similar assumption must be made to extrapolate the action spectrum for melanoma induction in fish to melanoma induction in humans.[18]

Conclusion

There is no question but that action spectra have been useful in the understanding of many light-dependent reactions ranging from photosynthesis to carcinogenesis. As with most techniques, however, there is nothing like understanding to make the technique effective. To quote Roderick Clayton (p. 107 of ref. 20)[20] "Historically the measurement of action spectra has provided much valuable information and also much avoidable confusion. When measured carefully and properly, the action spectrum is of greatest value in photobiology."

Acknowledgment

The author's work, reported in Figures 1.1 and 1.2, was supported by the Office of Health and Environmental Research of the U.S. Department of Energy.

References

1. Setlow R. Action spectroscopy. Adv Biol Med Phys 1957; 5:37-74.
2. Coohill TP. Action spectra again? Photochem Photobiol 1991; 54:859-870.
3. Morowitz HJ. Absorption effects in volume irradiation of microorganisms. Science 1950; 111:229-230.
4. Anderson RR. Tissue optics and photoimmunology. In: Parrish JA, Kripke ML, Morison WL, eds. Photoimmunology. New York:Plenum Med Bk Co., 1983:61-76.
5. Kelfkens G, DeGruijl FR, Van der Leun JC. Tumorigenesis by shortwave ultraviolet-A: Papillomas versus squamous cell carcinomas. Carcinogen 1991; 12:1377-1382.
6. DeFabo EC, Noonan FP. Mechanism of immune suppression by ultraviolet irradiation in vivo I. Evidence for the existence of a unique photoreceptor in skin and its role in photoimmunology. J Exp Med 1983; 158:84-98.
7. Gibbs NK, Norval M, Traynor NJ et al. Action spectra for the *trans* to *cis* isomerization of urocanic acid in vitro and in vivo. Photochem Photobiol 1993; 57:584-590.
8. Gibbs NK. Correction: Action spectra for the *trans* to *cis* isomerization of urocanic acid in vitro and in vivo. Photochem Photobiol 1993; 58:769.
9. Morrison H, Avnir D, Bernasconi C et al. Z/E photoisomerization of urocanic acid. Photochem Photobiol 1980; 32:711-714.
10. Morrison H, Bernasconi C, Pandey G. A wavelength effect on urocanic acid E/Z photoisomerization. Photochem Photobiol 1984; 40:549-550.
11. Kraemer KH, Lee ML, Scotto J. DNA repair protects against cutaneous and internal neoplasia: Evidence from xeroderma pigmentosum. Carcinogen 1984; 5:511-514.
12. Kraemer KH. Sunlight and skin cancer: Another link revealed. Proc Natl Acad Sci USA 1997; 94:11-14.
13. Setlow RB. The wavelengths in sunlight effective in producing skin cancer: A theoretical analysis. Proc Natl Acad Sci USA 1974; 71:3363-3366.
14. Freeman SE, Hacham H, Gange RW et al. Wavelength dependence of pyrimidine dimer formation in DNA of human skin irradiated in situ with ultraviolet light. Proc Natl Acad Sci USA 1989; 86:5605-5609.
15. Parrish JA, Jaenicke KF, Anderson RR. Erythema and melanogenesis action spectra of normal human skin. Photochem Photobiol 1982; 36:187-191.
16. Ley RD. Photoreactivation of UV-induced pyrimidine dimers and erythema in the marsupial *Monodelphis domestica*. Proc Natl Acad Sci USA 1985; 82:2409-2411.
17. DeGruijl FR, Sterenborg HJCM, Forbes PD et al. Wavelength dependence of skin cancer induction by ultraviolet irradiation of albino hairless mice. Cancer Res 1993; 53:53-60.

18. Setlow RB, Grist E, Thompson K et al. Wavelengths effective in the induction of malignant melanoma. Proc Natl Acad Sci USA 1993; 90:6666-6670.
19. DeGruijl FR, Van der Leun JC. Estimate of the wavelength dependency of ultraviolet carcinogenesis in humans and its relevance to risk assessment of stratospheric ozone depletion. Health Phys 1994; 67:319-325.
20. Clayton RK. Light and Living Matter, Volume 1: The Physical Part. New York:McGraw Hill, 1970:147.

CHAPTER 2

Sunscreens: The Molecules and Their Photochemistry

Bice S. Martincigh, John M. Allen and Sandra K. Allen

Introduction

The alarming worldwide increase in skin cancer is a major cause of concern and the center of much research. In the USA alone one million new cases of skin cancer are reported each year, of which 10,000 cases ultimately lead to death.[1] In countries like Australia and South Africa where the climate encourages the pursuit of outdoor activities, the incidence of skin cancer in the fair-skinned population is amongst the highest in the world.

The causes of skin cancer are complex, and in the case of malignant melanoma, are not well understood. However, for both malignant and nonmalignant melanoma, exposure to solar ultraviolet radiation has been implicated.[2]

A typical solar spectrum measured virtually at solar noon for a southern latitude of approximately 30° is shown in Figure 2.1. This spectrum of course changes with the solar zenith angle, the latitude of the location, the seasons and various atmospheric factors. Of the total solar energy reaching the Earth's surface, 15% is in the ultraviolet region, 60% is in the visible range and the remaining 25% is infrared radiation (or heat).[3] Despite the fact that ultraviolet radiation accounts for the smallest fraction of the terrestrial solar energy, this is the radiation with the shortest wavelengths and is consequently most energetic; it is therefore most likely to induce biological damage. Nevertheless, the possibility that other wavelengths may be able to incur damage cannot be excluded.

Solar ultraviolet radiation is divided into three regions according to the resultant biological effects. The ultraviolet C region (UVC) comprises the shortest wavelengths from 100 to 280 nm, but this region of the solar spectrum is absorbed mainly by ozone and by

Sunscreen Photobiology: Molecular, Cellular and Physiological Aspects, edited by Francis P. Gasparro.

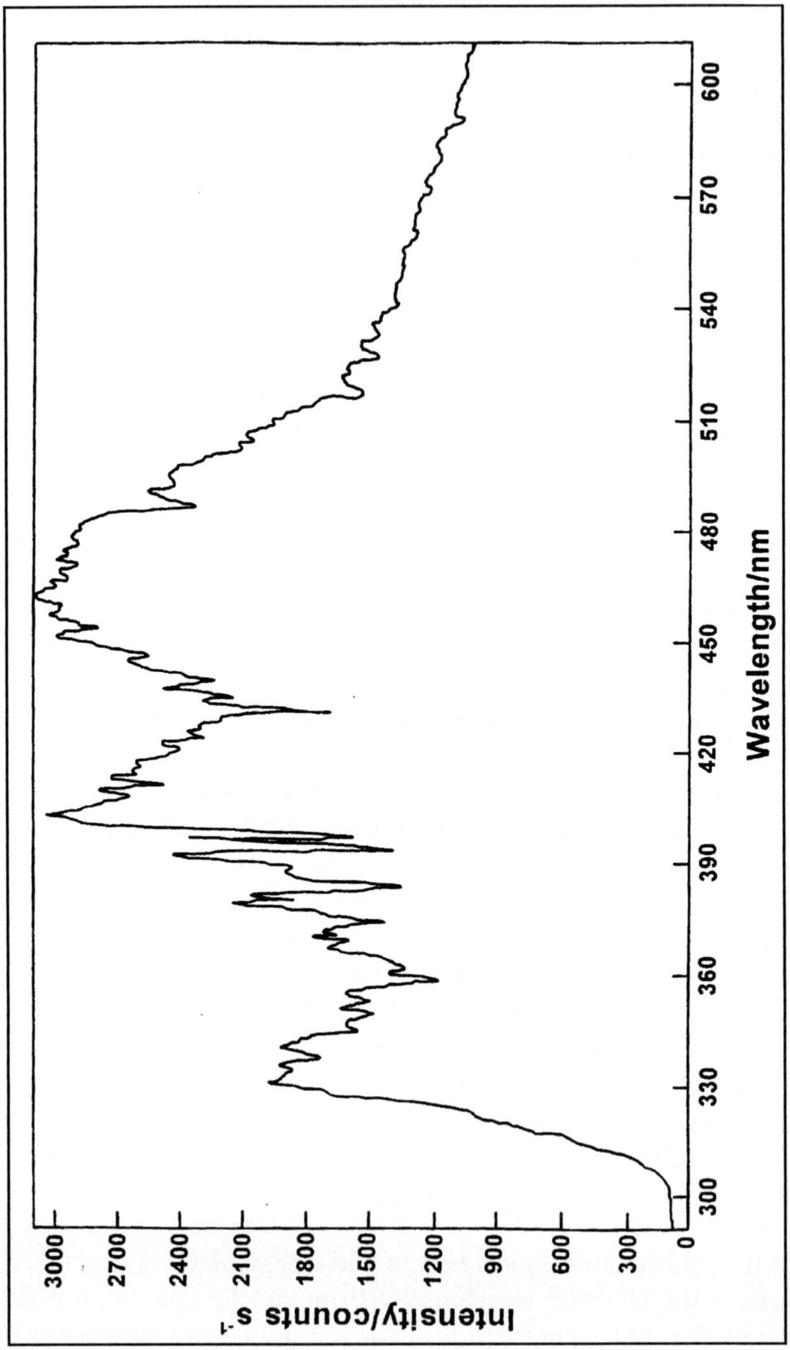

Fig. 2.1. A terrestrial solar spectrum recorded on 20 July 1995 at 12h10 UT in Durban, South Africa (29.9°S, 30.9°E). (Dr. E Mravlag of the Department of Physics, University of Natal, Durban, South Africa is thanked for providing these data.)

oxygen in the Earth's atmosphere and therefore does not reach the land surface.

The intermediate range is the ultraviolet B region (UVB) from 280 to 320 nm, a portion of which does not reach the Earth's surface because it is absorbed by stratospheric ozone. The shortest measurable solar wavelength at sea level is 290 nm. However, this wavelength cutoff is markedly dependent on the atmospheric ozone concentration and any depletion of the ozone layer will allow wavelengths shorter than this to penetrate to the terrestrial environment. As the absorption spectrum of DNA is spanned by the UVC and UVB, it is these wavelengths that are most destructive to living organisms, and under normal conditions of UV exposure, the UVB is most significant in the induction of skin cancers.[4] UVB is also the principal cause of erythema and it is implicated in structural and cellular skin damage such as elastosis, actinic keratosis and telangiectasis.[5]

The longer wavelength region, the ultraviolet A region (UVA), from 320 to 400 nm was once thought to be harmless because of its low energy, but is now known to cause wrinkling, photoaging of the skin, dermatological photosensitivity, some erythema, and has recently been shown to contribute to melanoma induction.[6] Recent work has also shown that on the molecular level UVA is capable of inducing single strand breaks and DNA-to-protein crosslinks in a manner similar to ionizing radiation.[7]

Of course, one's total exposure to the sun is also affected by one's occupation, clothing, lifestyle, etc. For example, a teacher is exposed to less solar ultraviolet radiation than a bricklayer, cotton clothes afford more protection than polyester ones, and a swimmer or golfer is exposed to more UV radiation than a table tennis player.

The coverage surrounding the depletion of stratospheric ozone and increased rates of skin cancer, has increased public awareness of the dangers of exposure to sunlight. As a response to this problem the use of sunscreen preparations to protect against the adverse effects of the sun is advocated. Consequently, the sunscreen industry is expanding rapidly and protective agents are now not only being incorporated into skin protection creams but also into almost every cosmetic product, including shampoos, moisturizers and lipsticks—all in the hope of affording the consumer more efficient protection.

The protective function of a sunscreen formulation is determined by its active ingredients. These are divided into two categories depending on their mode of action: chemical absorbers which absorb the sunburn-producing ultraviolet radiation, and physical blockers that reflect and scatter ultraviolet light. Lists of the sunscreen filters approved in various countries can be found in references 8-10. Table 2.1 gives the Category I UV absorbers approved by the U.S. Food and Drug Administration (FDA) for use in the USA. Included in Table 2.1 are the Chemical Abstract Service Registry

Table 2.1. List of U.S. FDA-approved sunscreen active ingredients

Name	CAS Number	Synonyms
Aminobenzoic acid	150-13-0	Benzoic acid, *p*-amino PABA Amben Aminobenzoic acid γ-Aminobenzoic acid *p*-Aminobenzoic acid *4*-Aminobenzoic acid *1*-Amino-*4*-carboxybenzene Anticanitic vitamin Anti-chromotrichia Factor Bacterial Vitamin H1 Benzoic acid, *4*-amino- *p*-Carboxyaniline *p*-Carboxyphenylamine Chromotrichia Factor anti-Chromotrichia Factor Kyselina *p*-aminobenzoova Pabanol Paraminol Paranate Sunbrella Trichochromogenic Factor Vitamin BX Vitamin H
Cinoxate	104-28-9	Propenoic acid, *3-(4*-methoxyphenyl)-, *2*-ethoxyethyl ester *2*-Ethoxyethyl *p*-methoxycinnamate Cinoxato (Spanish) Cinoxatum (Latin) Phiasol *2*-Propenoic acid, *3*-(*4*-methoxyphenyl)-, *2*-ethoxyethyl ester (9CI)

Name	CAS Number	Synonyms
Diethanolamine methoxycinnamate,	56265-46-4	Diethanolamine salt of *p*-methoxy hydrocinnamate *2-propionic acid, 3-{4*-methoxy phenyl)-, compound with 2,2'-iminobis[ethanol]
Digalloyl trioleate	526-08-3	Benzoic acid, *3,4*-dihydroxy-*5-5[(3,4,5*trihydroxybenzoyl)oxy]-
Dioxybenzone	131-53-3	Benzophenone, 2,2' *-dihydroxy-4-methoxy* Advastab 47 Benzophenone-8 Cyasorb UV 24 Cyasorb UV 24 Light Absorber*2,2' -Dihydroxy-4-methoxybenzophenone* Dioxybenzon Dioxybenzone Methanone, (2-hydroxy-4-methoxyphenyl) (2-hydroxyphenyl)-(9CI) Spectra-sorb UV 24 UF2 UV 24
Ethyl-4-[bis(hydroxypropyl)] aminobenzoate	58882-17-0	Ethyldihydroxy propyl PABA Ethyldihydroxypropyl p-aminobenzoate Benzoic acid, 4-[bis(2-hydroxyl propyl)amino]-ethyl ester
Glyceryl aminobenzoate	136-44-7	Glyceryl PABA *1,2,3-Propanetriol, p*-aminobenzoate *p*-Aminobenzoic acid mono-glyceryl ester Escalol 106 Glycerol, *1-p*-aminobenzoate Glycerol, *para*-aminobenzoate Monoglycerol, *p*-aminobenzoate *1,2,3*-Propanetriol, *1-(4*-aminobenzoate)
Homosalate	118-56-9	Benzoic acid, *2*-hydroxy-*3,3,5*-trimethylcyclohexyl ester (9CI) Caswell No. 482B Coppertone

Name	CAS Number	Synonyms
Homosalate (cont.)	118-56-9	EPA Pesticide Chemical Code 076603 Filtersol (A) Heliopan Heliophan Homosalato (Spanish) Homosalatum (Latin) Metahomomenthyl salicylate Salicylic acid, *m*-homomenthyl ester Salicylic acid, *3,3,5*-trimethylcyclohexyl ester (8CI) *3,3,5*-Trimethylcyclohexyl-*2*-hydroxybenzoate *3,3,5*-Trimethylcyclohexyl salicylate Uniderm Homsal
Lawsone	83-72-7	Lansone *1,4-Naphthalenedione, 2-hydroxyl*-C.I. 7 C.I. Natural Orange 6 Flower of Paradise Henna (dye) Henna *2-Hydroxy-1,4-naphthalenedione* *2-Hydroxy-1,4-naphthoquinone* *2-Hydroxynaphthoquinone* Mehendi Mendi
Dihydroxyacetone	96-26-4	Oxatone Chromelin *1,3-Dihydroxyacetone* *1,3-Dihydroxypropanone* Dihyxal Oxantin Soleal Triulose Viticolor
Methyl anthranilate	134-20-3	Benzoic acid, 2-amino-, methyl ester *2-Aminobenzoic acid methyl ester* *o-Aminobenzoic acid methyl ester*

Name	CAS Number	Synonyms
Methyl anthranilate (cont.)	134-20-3	Anthranilic acid, methyl ester *2-Carbomethoxyaniline* *o-Carbomethoxyaniline*
Methyl anthranilate (cont.)	134-20-3	*2-(Methoxycarbonyl') aniline* Methyl 2-aminobenzoate Methyl o-aminobenzoate
Octocrylene	6197-30-4	Octocriline *2-Ethylhexyl-2-cyano-3,3-diphenylacrylate* *2-Ethyhexyl-2-cyano-3,3-diphenyl-2*-propenoate Octocrileno (Spanish) *2-Propenoic acid, 2-cyano-3,3-diph*enyl-, 2-ethylhexyl ether (9CI) UV Absorber-3
Octyl methoxycinnamate	5466-77-3	2-Ethylhexyl methoxycinnamate 2-Ethylhexyl-4-methoxycinnamate 2-Ethylhexyl-*p*-methoxycinnamate 3-(4-Methoxyphenyl)-2-propenoic acid, 2-ethylhexyl ester (9CI)
Octyl salicylate	6969-49-9	Benzoic acid, 2-hydroxy octyl ester *n*-Octyl salicylate Salicylic acid, octyl ester
Oxybenzone	131-57-7	Benzophenone, 2-hydroxy-4-methoxy Benzophenone-3 Cyasorb UV 9 2-Hydroxy-4-methoxybenzophenone (2-Hydroxy-4-methoxyphenyl)-phenylmethanone Methanone, (2-hydroxy-4-methoxyphenyl)phenyl- *4-Methoxy-2-*hydroxybenzophenone MOB Oxybenzone

Name	CAS Number	Synonyms
Oxybenzone (cont.)	131-57-7	Spectra-Sorb UV9 Syntase 62 UF 3 USAF CY-9 Uvinul M 40
Padimate-O	21245-02-3	Octyl dimethyl PABA Arlatone UVB Benzoic acid, *4*-(dimethyl-amino)-, *2*-ethylhexyl ester (9CI) *4*-(Dimethylamino) benzoic acid, *2*-ethylhexyl ester Escalol 507 *2-Et*hylhexyl-*p*-dimethylaminobenzoate *2*-Ethylhexyl-*p*-(dimethyl-amino) benzoate Octyl dimethyl *p*-aminobenzoate Solarchem O UVasorb DMO
Phenylbenzimidazole sulfonic acid	27503-81-7	*2*-Phenyl Benzimidazole-*5*-sulphonic acid *2*-Phenylbenzimidazole-*5* sulfonic acid *1H*-Benzimidazole-*5*-sulfonic acid, *2*-phenyl-(9CI) *2-Phe*nyl-*1H*-benzimidazole-*5*-sulfonic acid
Red petrolatum	8009-03-8	Petrolatum
Sulisobenzone	4065-45-6	Benzenesulfonic acid, *5*-benzoyl-*4*-hydroxy-*2*-methoxy Benzophenone-4 *2*-Hydroxy-*4*-methoxybenzophenone-*5*-sulfonic acid
Titanium dioxide	1317-80-2	Titanium dioxide -Rutile
Trolamine salicylate	2174-16-5	Triethanolamine salicylate

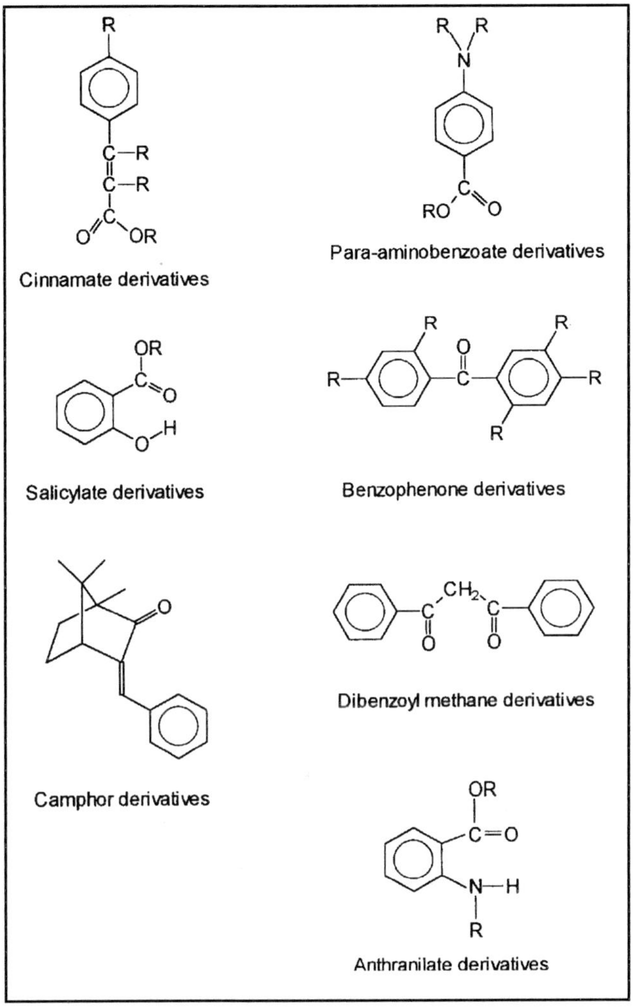

Fig. 2.2. The seven major groups of chemical sunscreen filters currently used in the suncare industry.

numbers (CAS numbers) and the various common names for each compound.

Chemical Absorbers

With the exception of a few miscellaneous compounds, the chemical absorbers currently used in the sunscreen industry can be classified into seven broad categories which are shown in Figure 2.2. In turn, these absorbers can be subdivided into either UVA (benzophenones, anthranilates and dibenzoyl methanes) or UVB (PABA derivatives, salicylates, cinnamates and camphor derivatives) filters depending upon which type of radiation they predominantly absorb.

In general these compounds usually contain an aromatic ring conjugated with a carbonyl group. Often an electron-releasing group such as an amine or methoxyl group, is substituted in the ortho- or para-position of the aromatic ring. In short therefore, these molecules contain conjugated systems which allow electron delocalization on absorption of a photon of UV light. They thereby achieve the purpose of absorbing any harmful UV radiation before it reaches the skin.

However, cognizance must be taken of the fact that any molecule with excess energy, and therefore in an excited state, is short-lived and must collapse back into its ground state. In doing so, various dissipative processes can occur which may be a source of damage to the skin. A brief overview of the absorption of light by organic molecules and the various possible pathways for the dissipation of the excess electronic energy (illustrated in Figure 2.3) follows.

On absorption of a photon of UV light, the energy of the molecule increases and the molecule attains a new electronic configuration. The amount of light absorbed at a particular wavelength for a given concentration of the substance is governed by the Beer-Lambert law:

$$A = \log \frac{I_o}{I} = \varepsilon bc$$

where A is the absorbance, I_0 is the intensity of the incident radiation, I is the intensity of the transmitted radiation, ε is the molar absorption coefficient (in units of $M^{-1}\ cm^{-1}$), b is the path length of the radiation through the sample in cm, and c is the concentration in molarity.

The ground state for most organic molecules is a singlet state in which all the electrons are paired with opposite spin, and at room temperature the lowest vibrational level ($\upsilon = 0$) of this singlet state is the one which is predominantly populated. In their excited states the molecules have two unpaired electrons in different orbitals. If the two electrons have opposite spins, the overall spin of the state is zero and this is referred to as a singlet state. On the other hand, if the two electrons have parallel spins, the overall spin is nonzero and the state is a triplet state. As the probability of spin inversion during

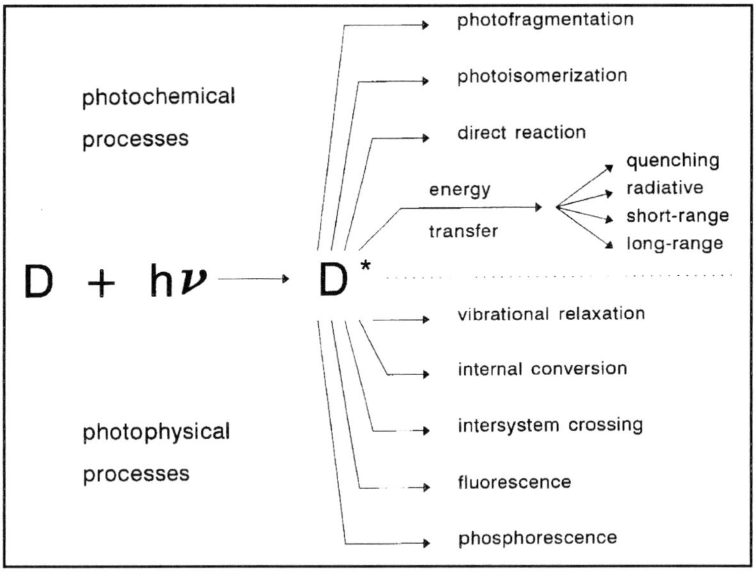

Fig. 2.3. An illustration of the absorption process and the various dissipative pathways available to an excited molecule D*.

the absorption process is low, the most probable transitions are those between states of the same spin.

Consequently, the most probable transition following the absorption of the photon is from the lowest vibrational level of the singlet ground state (S_0) to the first-excited singlet state (S_1). The molecule may now also, in addition to excess electronic energy, have excess vibrational and rotational energy. However, in situations where the rate of collisions is high (such as in solution), this excess vibrational energy is lost (vibrational relaxation) and is distributed as kinetic energy between the partners in the collision. Emission from S_1 is therefore most likely to occur from the lowest vibrational level.

Deactivation of an excited state can occur via various routes. These can be either photophysical or photochemical depending upon whether the resulting electronic state is of the same or a different molecule, respectively. The contribution made by each particular dissipative pathway is determined by the relative magnitude of the rate constant for that process. The inverse of the sum of all the first order (or pseudo first order) rate constants of the decay processes initiating from a particular excited state gives the lifetime of that excited state.

In the absence of any interacting chemical species, the photophysical processes leading to the loss of the excess energy of the excited molecule may be either radiative or radiationless transitions. The emission of radiation between two states of the same spin is known as fluorescence. This is a fast process and is usually observed from the lowest excited singlet state, S_1. Phosphorescence, on the other hand, is emission between two states of different spin and usually occurs from the lowest excited triplet state, T_1. As this is a "forbidden" transition (between two states of different spin), the emission occurs over a longer time period (smaller rate constant). As the triplet state is of lower energy than the corresponding singlet state, phosphorescence emission occurs at longer wavelengths than fluorescence emission.

Nonradiative processes are those that occur without emission of light between degenerate vibrational-rotational levels of different electronic states if the electronic states have the same spin. The transition is known as internal conversion (ic). When the transition occurs between states of different spin the process is called intersystem crossing (isc). The rates of both processes are inversely related to the energy difference between the lowest vibrational levels of the final and initial electronic states. Thus the smaller the energy gap, the higher the rate of the transition. As the energy separation between upper excited states is small, ic between these excited states is fast and accounts for the negligible emission of radiation from these states, whereas ic between S_1 and S_0 is slow because of the larger energy gap, therefore fluorescence can compete. Intersystem crossing from T_1 to S_0 competes with phosphorescence whereas the transition from S_1 to T_1 reduces the quantum yield of fluorescence and internal conversion to S_0, but is an important process for populating the triplet state. Therefore, for molecules with a small difference in energy between S_1 and T_1, isc is a fast and efficient process and after direct irradiation, reaction often proceeds from the lowest triplet state. This is the case with many ketones.

The various photophysical dissipative pathways described above can be illustrated on a Jablonski diagram such as that shown in Figure 2.4. In this diagram radiative transitions are represented by straight lines whereas nonradiative transitions are shown as wavy lines.

Because of the large energy separation between the lowest vibrational levels of S_1 and S_0 and of T_1 and S_0, ic and isc are relatively inefficient processes for an excited species to relax to the ground

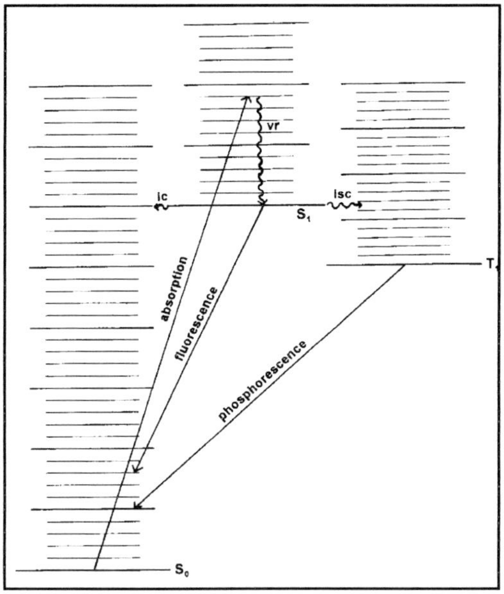

Fig. 2.4. A Jablonski diagram illustrating various intramolecular processes inititated by light absorption (ic - internal conversion, isc - intersystem crossing, vr - vibrational relaxation).

state S_0. Therefore the predominant pathways are either radiative emission or photochemical reaction, and in the absence of emission, it must be assumed that photochemical reaction is occurring. Chemical change can be brought about in a number of ways.

As the energy of the ultraviolet radiation which brings about electronic excitation is of the same order of magnitude as typical bond dissociation energies, the absorbing molecule can dissociate into reactive fragments (such as free radicals) or molecular species. Several mechanisms exist for the photofragmentation of the excited species: optical dissociation occurs immediately after the absorption step as the molecule is promoted into a dissociative vibrational level of the upper electronic state, whereas predissociation occurs after crossing from a stable vibrational level of one excited state to an unstable one of another excited state.

Photoexcitation of a molecule can also lead to direct reaction if the excited state produced is of sufficient energy for the molecule to cross the activation energy barrier. Consequently, the reaction is more likely to occur spontaneously than if it proceeded from the ground state. Thus a number of reactions which are not feasible at ambient temperatures can now take place.

As electronically excited species have considerably more energy than ground-state species, there is great potential for the occurrence of internal rearrangement of bonds and atomic positions.

This can ultimately lead to the formation of ground state products which are isomers of the initial absorbing molecule. In such cases the initial excited state formed relaxes to an excited state of lower energy but which does not retain the original geometry of the ground state (called a nonvertical excited state). This new state relaxes by radiationless decay to the ground states of the various possible isomers. Such a mechanism accounts for *cis-trans* photoisomerization.

Electronic energy transfer processes can cause deactivation of the initially-produced excited state. These are bimolecular interactions with the ground state of either another molecule or the selfsame species.

In the case where the energy transfer process results in deactivation of all excited species, D^*:

$$D^* + M \rightarrow D + M$$

the process is known as quenching. The energy initially absorbed by the nonexcited species M is converted into vibrational or translational excitation of M. If M happens to be a ground state molecule of D, the process is referred to as self-quenching.

Donation of the excess electronic energy to an appropriate acceptor molecule (A) is a common occurrence and an efficient means for deactivating excited states:

$$D^* + A \rightarrow D + A^*$$

Several mechanisms exist for the transfer of this excess energy depending upon the conditions involved. These mechanisms include: radiative transfer, short-range transfer, and long-range transfer.

In radiative transfer the radiation emitted by the donor (from either S_1 or T_1) is absorbed by the acceptor which is now excited:

$$D^* \rightarrow D + h\upsilon$$
$$A + h\upsilon \rightarrow A^*$$

In order for this to be an efficient process, the positions and intensities of the emission bands of the donor must match those of the absorption bands of the acceptor.

If the donor and acceptor can form a transient complex in which the separation distance between the two is within the collision diameter (but actual collision is not necessary), short-range energy transfer can occur. The energy states involved must be in accordance with the following spin requirements:

$$D^* (S_1) + A(S_0) \rightarrow D(S_0) + A^*(S_1)$$
$$D^* (T_1) + A(S_0) \rightarrow D(S_0) + A^*(T_1)$$

The former exchange is referred to as singlet-singlet energy transfer and the latter is triplet-triplet energy transfer. In both cases In order for the process to be efficient in both cases, the energy of D* must be greater than that of A^*.

Triplet-triplet energy transfer is an important route for producing triplet states which cannot easily be populated by direct absorption or where isc is inefficient. In such cases the transfer mechanism is referred to as photosensitization and the donor molecule is the photosensitizer. Aromatic ketones, such as benzophenone, which have very efficient isc from S_1 to T_1, can act as good photosensitizers.

Electronic energy transfer can still occur if the donor and acceptor are separated by a distance greater than the collision diameter. In this case the mechanism is referred to as long-range transfer or dipole-dipole mechanism. The efficiency of the process is dependent on the extent of overlap of the emission bands of the donor and the absorption bands of the acceptor. In this case, singlet-singlet energy transfer

$$D^* (S_1) + A(S_0) \rightarrow D(S_0) + A^*(S_1)$$

or triplet-singlet energy transfer

$$D^* (T_1) + A(S_0) \rightarrow D(S_0) + A^*(S_1)$$

can occur. Triplet-triplet energy transfer is unlikely because both processes are spin-forbidden.

A special mention must be made of the role of oxygen in these various dissipative pathways. The involvement of oxygen cannot be excluded as sunscreen absorbers are used in aerobic environments. Unlike most molecules that possess singlet ground states, oxygen has a triplet ground state (T_0) and a low energy singlet excited state (S_1). Consequently, it is a very good triplet quencher:

$$M^*(T_1) + O_2(T_0) \rightarrow M(S_0) + O_2^*(S_1).$$

The singlet oxygen produced is very reactive towards certain organic molecules. In addition, oxygen is a good electron acceptor and forms the superoxide radical anion $O_2^{\bullet-}$ in an electron-transfer reaction:

$M^*(S_1 \text{ or } T_1) + O_2(T_0) \rightarrow M^{\bullet+} + O_2^{\bullet-}$

The generation of $O_2^{\bullet-}$ in a biological system will produce hydrogen peroxide by dismutation:

$$2O_2^{\bullet-} + 2H^+ \rightarrow H_2O_2 + O_2$$

Although hydrogen peroxide, H_2O_2, is a weak oxidant and on its own not a great source of toxicity, it is a species of biological significance for the production of the extremely potent hydroxyl radical, $^{\bullet}OH$.[11] In the presence of free Fe^{2+} or $O_2^{\bullet-}$ hydrogen peroxide reacts as follows:

$$Fe^{2+} + H_2O_2 \rightarrow Fe^{3+} + HO^- + {}^{\bullet}OH \text{ (Fenton-type reaction)}$$
$$O_2^{\bullet-} + H_2O_2 \rightarrow O_2 + HO^- + {}^{\bullet}OH \text{ (Haber-Weiss reaction)}$$

Hydroxyl radicals are key species in producing toxic events.

Oxidation processes involving such reactive oxygen species are a well-known cause of cellular damage and cancer initiating events. The free radical species can induce peroxidation and polymerization in some of the skin components. Although these radicals play essential roles in aerobic life, any over-production can upset the balance, and toxicity and disease may result. Whether radicals act as toxic agents depends on their site of production, the biological relevance of their intracellular target, and the level of antioxidant activity.

These various dissipative processes for the relaxation of an excited molecule to the ground state will be illustrated in the case of chemical absorbers in the section entitled Photochemical Behavior of Chemical Absorbers. In addition, any potential deleterious consequences arising from the sunscreen photochemistry will be highlighted.

Physical Blockers

Current awareness of the serious skin damage that can be caused by UVA radiation has necessitated the introduction of suncare products that can offer broadband protection over both the UVA and UVB regions. The nature of chemical absorbers is such that they cannot span such a broad absorption range, so mixtures of these compounds must be used. Mixtures of absorbers also aid the formulator in producing high protection products without exceeding the regulatory concentration maxima set by various countries, and in overcoming problems of limited solubility of absorbers and incompatibility with other ingredients.

The limited number of suitable UVA absorbers has led the suncare industry to consider the use of physical blockers. These include such substances as titanium dioxide and zinc oxide. By decreasing the particle size of these oxides they become opaque to the UVA and UVB regions and transparent to the visible region. They are therefore attractive alternatives to chemical absorbers.

Titanium dioxide has been in use for many years as a whitener in paints, plastics, soap, paper and inks. In these applications the ability of titanium dioxide to impart an ultrawhite appearance to the finished products (or, in other words, opacity to visible light) is being exploited. This property is a consequence of the high refractive index of titanium dioxide of approximately 2.6[12] and the use of pigmentary size particles with a diameter of about 220 nm. Prior to using titanium dioxide, the whitener of choice was zinc oxide. However, zinc oxide has a lower refractive index of only 2.0[12] and does not impart as much whiteness and therefore was replaced by titanium dioxide.

These whitening properties are, however, unacceptable if the oxides are to be used in sunscreen preparations. How can this characteristic be reduced and still achieve a UV blocker? The answer lies in the mechanism of light attenuation for particles of different sizes.

For pigmentary particles the maximum opacity to visible light has been observed to occur from particle sizes ranging from 180 to 220 nm in diameter.[13] Light in this wavelength range is attenuated by Mie scattering in which the particles reflect and scatter the light and thereby prevent transmission. Maximum opacity is achieved by particles with a diameter which is approximately half the wavelength of the light. Particles much smaller than this merely transmit the light. Thus micronized oxides are transparent to visible radiation and can be used to produce aesthetically appealing cosmetic products.

For these smaller particle sizes the mechanism of light attenuation is explained by Rayleigh scattering. Here the intensity of the scattered light is inversely proportional to the fourth power of the wavelength. Thus light of short wavelengths is more efficiently scattered by small particles and reflection from the particle surface is minimal. How is the UV blocking achieved? The opacity in the UV region is achieved because these substances exhibit semiconductor properties and any light of energy greater than that of the band gap between the valence band and the conduction band is absorbed. Titanium dioxide has two crystalline forms: anatase and rutile, and their characteristic wavelengths corresponding to the band gap energy are 387 and 405 nm, respectively.[14] For zinc oxide the characteristic wavelength is 384 nm.[14] Thus for particle sizes between 20 and 100 nm UV radiation is effectively scattered and absorbed by the particles, thereby making these substances suitable UV blockers. Unfortunately, the maximum UVA and UVB opacities occur at different particle sizes.[5] However, it is possible to select an intermediate size which provides an acceptable level of UV opacity as well as cosmetic aestheticism.

Both titanium dioxide and zinc oxide have been regarded as inert and nontoxic: titanium dioxide through its use as a food and pharmaceutical color additive, and zinc oxide through its use in skin ointment creams. However, because of the semiconductor properties of both of these oxides they have been increasingly used as photocatalysts, particularly for the environmentally clean degradation of organic pollutants in wastewaters. Since, by their nature, sunscreen preparations are exposed to sunlight, the photocatalytic behavior of these physical blockers needs to be considered. This aspect will be elucidated further in the section entitled Photochemical Formation of Reactive Species by Physical Blockers.

Photochemical Behavior of Chemical Absorbers

Ideally, the energy acquired by a sunscreen active ingredient from absorbed radiation should be dissipated by radiative decay (fluorescence and phosphorescence), self-quenching or by internal conversion and vibrational relaxation (heat) as described above. Sunscreen active ingredients may also undergo photofragmentation or photoisomerization, transfer the energy acquired by absorption to other molecules, or when applied to skin, react directly with biomolecules.

Photofragmentation

The photofragmentation of sunscreen active ingredients is undesirable for two reasons: (1) it causes a loss of the active ingredient which can lead to a reduction in efficacy for the sunscreen preparation and (2) it can lead to the formation of short-lived reactive intermediates (e.g. free radicals) and to the formation of a variety of photoproducts. It is clearly undesirable for a significant loss of the sunscreen active ingredient to occur during the course of normal usage. This can result in the consumers of sunscreen products gaining an exaggerated sense of protection while experiencing a greater than expected exposure to solar UV radiation during the course of outdoor activities. The formation of free radicals and other short-lived reactive intermediates in illuminated solutions of sunscreen active ingredients has previously been reported in the literature.[15-17] The toxicity of free radicals and other reactive species is well established.[18] Some of the stable photoproducts derived from sunscreen active ingredients may also be toxic. What is not clear at this point is whether or not the formation of reactive intermediates and photoproducts poses a significant risk factor to health under actual sunscreen use conditions.

Consider a sunscreen active ingredient molecule represented by RC-CR' where R and R' are organic groups. Given the appropriate R groups, this molecule can absorb UVB or UVA electromagnetic radiation. It must be kept in mind that sunscreen active ingredients have been specifically designed to absorb solar photons that have enough energy to break many covalent bonds. Thus, when the energy absorbed is great enough to rupture one or more covalent bonds in the molecule and the competing pathways for energy dissipation such as fluorescence are inefficient, photofragmentation can occur:

$$\mathrm{RC\text{-}CR'} + h\nu \rightarrow \mathrm{RC\text{-}CR'^{*}}$$
$$\mathrm{RC\text{-}CR'^{*}} \rightarrow \mathrm{RC^{\bullet}} + {}^{\bullet}\mathrm{CR'}$$

where $RC^{\bullet}$ and ${}^{\bullet}CR'$ represent carbon-centered free radicals. These free radical species are very reactive and can attack and modify other molecules in the immediate vicinity, yielding a variety of free radicals where the R group is widely variable depending upon the identity of the molecules that are available for reaction. This ultimately leads to a suite of stable photoproducts via free radical termination reactions. The identity of these photoproducts is dependent upon the chemical environment in which photofragmentation takes place.

Under the conditions of normal sunscreen usage, molecular oxygen (O_2) is present in the sunscreen preparation as the result of dissolution from the surrounding air. The free radicals formed as a result of photofragmentation can react with molecular oxygen in the following manner:

$$RC^{\bullet} + O_2 \rightarrow RCOO^{\bullet}$$

where $RCOO^{\bullet}$ is an organic peroxyl radical. Organic peroxyl radicals can react with a variety of organic molecules mainly via hydrogen atom abstraction and addition across unsaturated bonds. Reactions with the resonance-weakened O-H, S-H, and N-H functional groups present in phenols, thiophenols, and aromatic amines are particularly rapid ($k \approx 10^9\ M^{-1}\ s^{-1}$)[19] and peroxidation reactions with cellular lipids lead to the formation of hydroperoxide products:[18]

$$RCOO^{\bullet} + R'H \rightarrow RCOOH + R'^{\bullet}$$

where R'H represents a cellular lipid and RCOOH represents a lipid hydroperoxide product. The carbon-centered radical $R'^{\bullet}$ also formed can then react with oxygen, yielding another peroxyl radical. Termination self reactions involving peroxyl radicals yield dialkylperoxides (RCOOCR) and (O_2):

$$2RCOO^{\bullet} \rightarrow [RCOOOOCR] \rightarrow RCOOCR + O_2$$

The dialkylperoxide products can break down further to yield free radicals. Singlet molecular oxygen is discussed in some detail in section 2.4.

Alternatively, photofragmentation can lead to the formation of other reactive species via the following mechanism:

$$RC\text{-}CR' + h\upsilon \rightarrow RC\text{-}CR'^{*}$$
$$RC\text{-}CR'^{*} \rightarrow RC\text{-}CR'^{\bullet +} + e^{-}_{solvated}$$

where $RC\text{-}CR'^{+}$ is a radical cation derived from the sunscreen molecule. Formation of solvated electron ($e^{-}_{solvated}$) is significant because it can rapidly reduce molecular oxygen ($k \approx 10^{10}\ M^{-1}\ s^{-1}$), leading to formation of superoxide radical anion ($O_2^{\bullet -}$). Superoxide radical anion is in an acid-base equilibrium with hydroperoxyl radical ($HOO^{\bullet}$):

$$O_2^{\bullet -} + H^{+} \quad HOO^{\bullet}$$

The pK_a for $HOO^{\bullet}$ is 4.8. Superoxide radical anion is a selective oxidant and is relatively unreactive toward most biomolecules. However, the so-called dismutation (disproportionation) reactions involving $O_2^{\bullet-}$ and $HOO^{\bullet}$ can lead to rapid formation of hydrogen peroxide:[20]

$$HOO^{\bullet} + O_2^{\bullet-} + H^+ \rightarrow H_2O_2 + O_2 \qquad k_2 = 1 \times 10^8\ M^{-1}\ s^{-1}$$
$$HOO^{\bullet} + HOO^{\bullet} \rightarrow H_2O_2 + O_2 \qquad k_2 = 8.6 \times 10^5\ M^{-1}\ s^{-1}$$
$$O_2^{\bullet-} + O_2^{\bullet-} + 2H_2O \rightarrow$$
$$H_2O_2 + O_2 + 2^{\bullet}OH \qquad k < -0.3\ M^{-1}\ s^{-1}$$

At pH = 7, the dismutation equilibrium lies far to the right:

$$2O_2^{\bullet-} + 2H^+ \rightarrow H_2O_2 + O_2 \qquad K = 4 \times 10^{20}$$

Even at pH = 14, dismutation is complete ($K = 9.1 \times 10^8$). It is likely that some of the oxygen molecules formed in the dismutation reactions above are present as $^1O_2$.

The significance of hydrogen peroxide formation is 2-fold: (1) it is an oxidant that can react with a variety of biomolecules, thereby exerting a toxic effect; and (2) as discussed previously, it serves as a source for the very powerful oxidant, hydroxyl radical ($^{\bullet}OH$). Hydrogen peroxide can react with certain biomolecules forming hydroperoxide products which can decompose, ultimately yielding free radicals. In the presence of reduced transition metals such as Fe^{2+}, which is present at variable concentrations in biological systems as well as in sunscreen formulations, H_2O_2 may react via Fenton-type reactions to form $^{\bullet}OH$:

$$H_2O_2 + Fe^{2+} \rightarrow FeOH^{2+} + {}^{\bullet}OH$$

An alternative source of $^{\bullet}OH$ is via Haber-Weiss reactions:

$$O_2^{\bullet-} + H_2O_2 \rightarrow {}^{\bullet}OH + O_2 + OH^-$$

Hydrogen peroxide can also undergo direct photofragmentation yielding $^{\bullet}OH$:

$$H_2O_2 + h\nu \rightarrow 2^{\bullet}OH$$

Photofragmentation of H_2O_2 in sunlight is relatively slow.

Hydroxyl radical reacts with organic compounds via hydrogen atom abstraction, addition, and electron transfer mechanisms. Its reactions with a variety of organic and biomolecules occur at nearly diffusion-controlled ($k \approx 10^{10}\ M^{-1}\ s^{-1}$) rates, forming different (e.g. carbon-centered and the corresponding peroxyl) free radicals. For example, $^{\bullet}OH$ reacts with thymine by addition across a double bond yielding thymine radical and, in the presence of oxygen, thymine peroxyl radical. These reactions can lead to DNA damage via strand breakage, cell mutation, and ultimately, cell death.

It is important to remember that under continuous illumination conditions such as in sunlight, free radicals are formed continuously and are rapidly consumed by reactions with organic and biological molecules, other free radicals, and molecular oxygen. The result of this is that a steady-state is rapidly reached in which the formation rates for a large number of free radicals is equivalent to the rates of loss for these radicals. The steady-state concentrations of these reactive intermediates are likely to be quite low in sunlight illuminated sunscreen formulations. Other reactive species such as H_2O_2 and $^1O_2$ are continuously formed but react more selectively than free radicals such as $^{\bullet}OH$. That they are likely therefore to reach significantly higher (several orders of magnitude) steady-state concentrations.

Few experimental studies involving the photochemistry of sunscreen active ingredients have been published in the literature. None of the published studies have provided data that allows an estimate to be made of the rate at which sunscreen active ingredients degrade in sunlight by photofragmentation or photoisomerization. This is because published stability studies have either used light sources that do not accurately simulate natural sunlight or they have used a method for measuring the rate of loss for sunscreen active ingredients (e.g. optical absorbance) that leads to ambiguous results.

The selection of illumination sources used for sunscreen photochemical studies is crucial. It is desirable to use natural sunlight for these studies when possible. In practice, it is not always convenient to use natural sunlight. A wide variety of illumination sources are commercially available. At present, the best available illumination sources for solar simulation are based upon filtered xenon arc lamps. The construction of such solar simulator illumination sources has been described in detail.[21]

Measurement of the change in optical absorbance at a specific wavelength (typically λ_{max}), or measurement of the change in the optical absorption spectrum as a function of illumination time, can lead to erroneous conclusions about the rate of photofragmentation and photoisomerization (i.e. photostability) because photoproducts may absorb at the same wavelength as the sunscreen itself. In practice, this means that a sunscreen active ingredient may be completely converted to a photoproduct or products and yet be judged to have a high degree of photostability because it has retained its original absorption spectrum after a period of illumination. Absorbance versus illumination time measurements are useful, nevertheless, because such measurements allow an estimation of the loss of absorbance, and therefore efficacy, as a result of a specific illumination exposure.

When conducting photostability studies, it is preferable to use a separation method such as gas chromatography (GC) or high performance liquid chromatography (HPLC) in order to separate the sunscreen active ingredient from any photoproducts and thus follow the loss of the sunscreen active ingredient without interference as a function of illumination time.

Several reported studies of sunscreen photochemistry have centered upon identification of reactive intermediates and stable photoproducts derived from *p*-aminobenzoic acid (PABA). In one of the first of these studies,[15] EPR spectroscopy was used to identify reactive intermediates formed upon photofragmentation of PABA. It was reported that illumination of PABA using the Pyrex-filtered output from a xenon arc lamp yielded benzoic acid radicals, and solvated electrons. It has been reported[22] that *cis*- and *trans*-4,4′-azodibenzoic acid were formed when PABA was illuminated with 313 nm light in aqueous solution. Shaw et al[23] reported that 4-amino-3-hydroxybenzoic acid, 4-aminophenol, and 4-(4′-hydroxyphenyl) aminobenzoic acid were obtained when aqueous solutions containing PABA were illuminated using Pyrex filtered UV (Rayonet-3000) lamps. Both of these publications[22,23] proposed mechanisms for the formation of observed products that involved free radical intermediates.

Reports of the photochemistry of other sunscreen active ingredients are less numerous. The photochemistry of Parsol-1789 (1-[4-(1,1-dimethylethyl)phenyl]-3-(4-methoxyphenyl)-1,3-propanedione), Oxybenzone ((2-hydroxy-4-methoxyphenyl)phenylmethanone), and Padimate-O (2-ethylhexyl-4-(dimethylamino)benzoate) were investigated and reported by Roscher et al.[24] Illumination of Parsol-1789 in cyclohexane solutions using an unfiltered mercury lamp yielded *tert*-butylbenzene, *p*-*tert*-butylbenzoic acid, and *p*-methoxybenzoic acid. Oxybenzone did not photodegrade. Photofragmentation of

Padimate-O yielded ethylhexyl esters of *p*-aminobenzoic acid, *p*-methylaminobenzoic acid, and *p*-dimethylaminomethylbenzoic acid. Indirect evidence for the formation of $^{\bullet}OH$ in illuminated solutions containing Padimate-O has also been reported.[25] The oxidation of Oxybenzone after topical skin application was followed by Fourier transform Raman spectroscopy.[26] The authors reported that Oxybenzone semiquinone was formed via this reaction.

Direct Reaction

If sunscreening agents are able to penetrate the skin, the possibility exists for the occurrence of UV light-induced reactions between the sunscreen molecules and the biomolecules in their vicinity. Besides oxidative damage, any sunscreen-DNA photoadducts formed could be potentially carcinogenic as they could block the reading of the genetic code and cause errors in DNA replication.

Gasparro and Battista[27] reported that PABA could bind covalently to various homopolynucleotides of the DNA bases, as well as to DNA itself. More recently, Shaw et al[28] have investigated the photochemistry of PABA with both thymine and thymidine in order to elucidate the mechanism of PABA photobinding to DNA components. They were able to isolate and characterize thymine-PABA and thymidine-PABA photoadducts.

There has also been speculation[29] that 2-ethylhexyl-*para*-methoxycinnamate (2-EHMC) may be able to photobind to DNA. Mohammad et al[30] reported that on UV irradition at wavelengths of 266 and 308 nm, radiolabeled *para*-methoxycinnamic acid becomes incorporated into calf thymus DNA. The radiolabel was released on photolysis at 254 nm, which suggests that binding occurred through [2 + 2] cyclobutane adduct formation. Since 2-EHMC is essentially similar in structure to *para*-methoxycinnamic acid (apart from the 2-ethylhexyl chain which replaces the carboxylic acid group) and has been shown to bind with itself,[29] there is no reason to suppose that it will not bind similarly to DNA components.

Photoisomerization

The cinnamate absorbers are subject to *trans-cis* photoisomerization across the ethylenic double bond. This isomerization is accompanied by a decrease in the absorbance maximum as the *cis*-isomer is a less efficient UV absorber. If, in addition to the isomerization reaction, nonabsorbing degradation products are produced, a further loss of screening efficiency is observed. As mentioned previously, this loss of photoprotection in a sunscreen absorber is of particular concern as is the build-up of possibly phototoxic products on the skin.

Trans-2-ethylhexyl-*para*-methoxycinnamate, which is currently the most commonly used UVB absorber in sunscreen preparations, shows a loss in absorbing ability[16] on exposure to UV light as a result of isomerization to the *cis*-isomer. Several investigators[16,31-37] have studied the absorption loss and reported losses ranging from 4.5% to 50%. These conflicting results are probably due to the dissimilar conditions and irradiation lamps used. The reasons for the absorption loss have been ascribed to photoisomerization only, photodegradation only, or to a combination of both processes. Broadbent et al[29] have shown that when neat material is irradiated with wavelengths greater than 300 nm, *trans*-2-EHMC not only photoisomerizes to the *cis*-isomer but also dimerizes with itself by means of a [2 + 2] cycloaddition reaction across the ethylenic double bond.

Another class of UV absorbers which undergo photoisomerization reactions are the dibenzoyl methane derivatives. These exhibit keto/enol tautomerism. Deflandre et al[36] found that this photoreaction caused a significant loss in absorption and hence protective power, in those derivatives in which a hydroxyl group was not present in the *ortho*-position to the carbonyl group.

Energy Transfer

It is also possible for chemical sunscreen absorbers to act as photosensitizers. This process, which involves energy transfer to another molecule, competes with the other dissipative processes outlined above. In some cases, this is a desirable mechanism for energy dissipation. In other cases, it leads to the formation of undesirable products. Singlet molecular oxygen ($^1O_2$), a potentially toxic oxidant can be formed by such an energy transfer process. Singlet molecular oxygen can be readily formed by the absorption of sunlight-range photons by photosensitizers such as some of the sunscreen absorbers (D):

$$D(S_1) + h\upsilon \rightarrow D^*(S_1)$$
$$D^*(S_1) \rightarrow D^*(T_1)$$
$$D^*(T_1) + O_2(^3\Sigma_g) \rightarrow O_2(^1\Delta_g) + D(S_1)$$

where $O_2(^3\Sigma_g)$ is ground state triplet molecular oxygen and $O_2^*(^1\Delta_g)$ is electronically excited singlet molecular oxygen, designated herein simply as $^1O_2$. The photosensitizer, initially in a singlet ground electronic state, is promoted by the absorption of a photon to an excited singlet state and can subsequently undergo intersystem crossing to form an excited triplet, which can then promote triplet ground state oxygen to an excited singlet state. Singlet molecular oxygen is a significantly more effective oxidant for most organic molecules than ground state molecular oxygen because its interactions with singlet ground state organic molecules are spin allowed.

Many organic chemical compounds have been shown to function as photosensitizers and thus produce $^1O_2$ when illuminated in air-saturated solution. The energy required to promote molecular oxygen from the ground-state triplet to the excited singlet state is only 95 kJ mol^{-1}.[38] The energy of the triplet excited state of PABA has been reported as 315 kJ mol^{-1} by Gonzenbach et al[39] and as 297 kJ mol^{-1} by Halpern and Ramachandran.[40] Thus, this excited state triplet of PABA clearly has enough energy to promote $O_2(^3\Sigma_g)$ to $^1O_2$. The triplet energies of 4-*tert*-butyl-4′-methoxydibenzoylmethane and 2-ethylhexyl-*p*-methoxycinnamate have been estimated by Gonzenbach et al[39] to be 249 and 238 kJ mol^{-1}, respectively. Thus, the triplet energies for all of these compounds are high enough to form $^1O_2$. The triplet energies of other sunscreen active ingredients have not been reported in the literature.

It has been demonstrated by Allen et al[17,41] that $^1O_2$ is formed when several sunscreen active ingredients are illuminated. Individual aqueous solutions containing a single sunscreen active ingredient and a chemical trap for $^1O_2$ were continuously illuminated in a xenon arc solar simulator. The authors found that $^1O_2$ is formed in illuminated aqueous solutions containing PABA, Padimate-O, 2-ethylhexyl-*p*-methoxycinnamate, and Octocrylene (2-ethylhexyl-2-cyano-3,3-diphenylacrylate). The $^1O_2$ formation rate in illuminated PABA solutions was found to be significantly higher than for all of the other evaluated compounds. The $^1O_2$ concentration when individual solutions containing Octylsalicylate (2-ethylhexylsalicylate), Benzophenone-3 (2-hydroxy-4-methoxybenzophenone) and Benzophenone-8 (2,2′-dihydroxy-4-methoxybenzophenone) were illuminated was below the experimental limit of detection. There was also no evidence found for the formation of any other reactive free radical intermediates (e.g. hydroxyl and peroxyl radicals) in illuminated solutions containing Octylsalicylate, Benzophenone-3 and Benzophenone-8.

Another important aspect of energy transfer in sunscreen active ingredients is the potential for DNA base damage. Epidemiological and clinical studies have shown that the UVB portion of the solar spectrum is primarily responsible for the induction of nonmelanoma skin cancer. This radiation is readily absorbed by biological molecules such as proteins and DNA, and is most effective in inducing DNA lesions in human skin. If left unrepaired these lesions alter the biological function of DNA and cause lethal, mutagenic and carcinogenic effects.

The photoreactive species in DNA are the bases and the pyrimidine bases, thymine and cytosine in particular. These bases absorb the ultraviolet light and are excited into higher energy electronic states which allows reactions to occur with adjacent bases on the DNA strand. The most abundant photoproducts are the cyclobutane pyrimidine dimers. Of these the thymine-thymine cyclobutane dimer forms predominantly followed by thymine-cytosine and cytosine-cytosine dimers. Other photoproducts which have been identified are the (6-4) pyrimidine-pyrimidone photoadducts and the pyrimidine photohydrates. These lesions are thought to be precursors to skin cancer. Cellular repair mechanisms can remove such lesions and prevent tumor formation. It is also known that ultraviolet light decreases the effectiveness of the immune system.

The yields of lesions are greatest when formed with light of about 260 nm (which is filtered out by ozone) and decrease with increasing wavelength. Significant amounts of the pyrimidine dimers can, however, be formed at longer wavelengths in the presence of triplet state photosensitizers. The only requirement for this reaction to occur is that the triplet energy level of the excited base be lower than that of the photosensitizer molecule. Photosensitizers which have been shown to induce damage in DNA include benzophenone, acetophenone and acetone.[42] It has also been shown that certain chemical absorbers used in sunscreen preparations to block UVB (and thus prevent erythema) can act as such photosensitizers and induce damage. Salter and co-workers have investigated the ability of various chemical absorbers to photosensitize the formation of thymine dimer in DNA and related systems.[43-46] Their research has focused on understanding the kinetic and mechanistic details of the

Table 2.2. Mechanism for PABA photosensitized formation of thymine dimer in pUC19 plasmid DNA at pH 7.[46]

Reaction	No.	Rate constant
$P \rightarrow {}^1P$	(1)	$4.795 \times 10^{-5}\ s^{-1}$
${}^1P \rightarrow P$	(2)	$1.27 \times 10^{9}\ s^{-1}$
${}^1P \rightarrow P$	(3)	$1.49 \times 10^{10}\ s^{-1}$
${}^1P \rightarrow {}^3P$	(4)	$1.22 \times 10^{9}\ s^{-1}$
${}^3P \rightarrow P$	(5)	$1.76 \times 10^{5}\ s^{-1}$
${}^3P + P \rightarrow 2P$	(6)	$2.94 \times 10^{9}\ M^{-1}\ s^{-1}$
${}^3P + X \rightarrow P + X$	(7)	$2.89 \times 10^{9}\ M^{-1}\ s^{-1}$
${}^3P + TcT \rightarrow P + {}^3TcT$	(8)	$3.7 \times 10^{9}\ M^{-1}\ s^{-1}$
${}^3TcT \rightarrow TcT$	(9)	$8.12 \times 10^{7}\ s^{-1}$
${}^3TcT \rightarrow T<>T$	(10)	$1.39 \times 10^{8}\ s^{-1}$

P = ground state PABA
^{1}P = singlet state PABA
^{3}P = triplet state PABA
TcT = ground state contiguous thymine
^{3}TcT = triplet state contiguous thymine
T<>T = thymine dimer
X = all non-contiguous thymine moieties in DNA

Table 2.3. Chemical sunscreen absorbers and some of their derivatives, which are able to photosensitize the formation of thymine cyclobutane dimers in aqueous solutions of free thymine.

p-aminobenzoic acid[45]
n-ethoxylated-p-aminobenzoic acid ethyl ester[49]
n-butyl-p-aminobenzoate[49]
n-ethyl-p-aminobenzoate[49]
n-butyl-p-diethyl-aminobenzoate[49]
disodium-2,2'-dihydroxy-4,4'-dimethoxybenzophenone sulphonic acid[44]
2-phenylbenzimidazole-5-sulphonic acid[50]

photosensitized pyrimidine dimerization process in DNA. The in vitro PABA photosensitized formation of thymine dimer in pUC19 plasmid DNA will be described as an illustration.[46] This system is particularly relevant to in vivo DNA since PABA has been shown to potentiate thymine dimer formation in cells.[47]

The mechanism proposed[46] for the PABA photosensitization of pUC19 DNA is shown in Table 2.2.

PABA was assumed to be the sole absorber in these experiments. On absorbing a photon of incident light the electrons in the PABA molecule rearrange into a more energetic singlet state (^1P) (reaction 1). This singlet state is short-lived and rapidly intersystem crosses to the triplet state (^3P) (reaction 4). The ^3P can then (by diffusion) collide with a contiguous thymine pair on the DNA strand to form a triplet thymine pair (^3TcT) (reaction 8) which then forms a dimer (T<>T) (reaction 10). Other steps are included in the mechanism to account for the quenching of ^3P by various species, or the loss of its energy by radiationless transition.

On collision with DNA the PABA triplet has only enough energy to populate the thymine triplet state so products derived from the other bases or the 6-4 photoproduct are not formed. The formation of thymine dimer by the direct irradiation of DNA at wavelengths greater than 300 nm is very small when compared to the yields produced via the PABA sensitization. Hence this source of thymine dimer was neglected.

The validity of the above mechanism was tested by using it to calculate the expected thymine dimer yields for the various experimental conditions investigated. The results obtained were compared with the experimentally obtained quantities and fair agreement was achieved. Discrepancies were observed for periods of long irradiation. This is understandable because in such cases significant yields of various PABA and thymine-derived photoproducts[48] which were not accounted for in the mechanism, were obtained. Reactions for the formation of these photoproducts were not included in the above mechanism because this would require knowledge of the absorption and photosensitizing properties of the photoproducts. The formation of these unknown photoproducts indicates that the potential of PABA to cause damage to DNA is not limited to the photosensitization of thymine dimer formation. It seems therefore that PABA is not suitable as a sunscreen constituent.

Table 2.3 lists the various chemical absorbers which have been found to photosensitize the formation of thymine dimer in aqueous solutions of free thymine base.

Mechanisms such as the one detailed above suggest the qualities a good absorber in a sunscreen should possess so as not to exhibit these potentially mutagenic effects. It should absorb UVB radiation effectively but be quickly converted to a harmless form, possibly by quenching with itself. Otherwise it should not be able to act as a triplet state photosensitizer, either by having a low rate of isc or by having a triplet energy that is sufficiently lower than that of the DNA base with the lowest triplet energy, namely thymine. For example, 2-EHMC is unable to photosensitize pyrimidine dimerization as its triplet state is sufficiently below that of thymine. It is therefore unable to excite thymine to its triplet state which is the precursor to the thymine dimer.

Photochemical Formation of Reactive Species by Physical Blockers

As discussed previously, titanium dioxide and the other metal oxides are semiconductor materials that have been used extensively for a variety of purposes, including use as photocatalysts for destruction of liquid hazardous waste. Upon absorption of radiation by TiO_2, an electron is promoted from the valence band (VB) to the conduction band (CB) leaving a hole (h^+) in the valence band:[51]

$$TiO_2 + h\upsilon \rightarrow h_{VB}^{+} + e_{CB}^{-}$$
$$h_{VB}^{+} + {>}TiOH \rightarrow\rightarrow {}^{\bullet}OH$$
$$h_{VB}^{+} + R \rightarrow R^{\bullet +}$$
$$e_{CB}^{-} + O_2 \rightarrow O_2^{\bullet -}$$
$$O_2^{\bullet -} + O_2^{\bullet -} + 2H^{+} \rightarrow H_2O_2 + O_2$$

where >TiOH is the hydrated TiO_2 surface, R is an organic compound adsorbed to the surface of the TiO_2 particle, and $R^{+\bullet}$ is an organic radical cation. Thus, absorption of a photon leads to the formation of a variety of reactive species including $^{\bullet}OH$, $R^{+\bullet}$, $O_2^{1\bullet}$/$HOO^{\bullet}$, and H_2O_2.

Titanium dioxide in micronized form is frequently present in sunscreen formulations along with organic active ingredients in order to provide a high sunscreen protection factor (SPF). Of current concern, is the potential for TiO_2 to form enough $^{\bullet}OH$ to significantly degrade the organic active ingredients in sunscreens when both TiO_2 and organic absorbers are present in mixed formulations. As was mentioned previously, $^{\bullet}OH$ is a very powerful oxidant that can react rapidly with virtually any organic compound, including sunscreen active ingredients. Reaction with $^{\bullet}OH$ could potentially result in the rapid loss of organic sunscreen active ingredients and to a potential loss in efficacy.

Another issue associated with the use of TiO_2 and other semiconductor sunblockers is the potential for formation of enough reactive oxygen species, such as $^{\bullet}OH$, on the skin to pose a significant risk of toxic effects. Of particular concern in this regard, is the fairly recent demonstration that photoexcited micronized TiO_2 can cause cell death both in vitro[52] and in vivo.[53] When these experiments were repeated in the presence of the $^{\bullet}OH$ quencher, L-tryptophan, and the H_2O_2 scavenger, catalase, cell death was significantly reduced, thereby suggesting that the cytotoxicity was due to the reactive oxygen species, $^{\bullet}OH$ and/or H_2O_2, produced by the photoexcited TiO_2. In addition, by means of transmission electron microscopy, TiO_2 was found to be present both in the cell membrane and the cytoplasm, but not in the nucleus, of the cells. If UVB radiation reaches a TiO_2 particle within a living cell, the potential exists for formation of $^{\bullet}OH$ and other reactive intermediates in the immediate vicinity of biomolecules including DNA, thereby imposing oxidative stress. Manufacturers coat the surfaces of the particles in an effort to reduce the photocatalytic activity of these metal oxide sunscreen. The effectiveness of these proprietary surface coatings has not yet been established.

Additional Considerations

As discussed previously, sunscreen absorbers can undergo photofragmentation and photoisomerization and physical blockers can act as photocatalysts, leading to the formation of reactive intermediates and stable photoproducts. However, the location at which formation of reactive intermediates and stable photoproducts takes place is extremely important. When very short-lived intermediates such as $^{\bullet}OH$ are formed in the stratum corneum, there is little chance to cause damage to viable cells. If however, $^{\bullet}OH$ is formed *inside* or in the immediate vicinity of living cells, they may well exert a significant toxic effect. Although Sutherland[47] has demonstrated that PABA can penetrate human cells, and TiO_2 is known to be transported into cells, as discussed previously, it is unclear to what degree sunscreen active ingredients are transported below the stratum corneum and into the epidermis and dermis after application. A significant fraction of incident solar UVA and, to a somewhat lesser degree, UVB radiation is transmitted through the unprotected stratum corneum and the epidermis to the dermis.[54] Further, in the case of H_2O_2 and $^1O_2$ which have significantly longer lifetimes than free radicals such as $^{\bullet}OH$, the picture is more complex. These species may be formed in the stratum corneum or epidermis and be transported to the dermis where they can do cellular damage.

Another important consideration is that sunscreen active ingredients are in some cases themselves scavengers of reactive intermediates. It has been suggested that sunscreen active ingredients exert protective effects by rapidly reacting with (scavenging) reactive intermediates such as free radicals. Many sunscreen active ingredients undoubtedly act as $^{\bullet}OH$ scavengers. It has also been suggested[55] and demonstrated experimentally[56] that PABA reacts very rapidly with $^1O_2$ ($k = 8.9 \times 10^8\ M^{-1}\ s^{-1}$). Hue et al[57] demonstrated that PABA reacts rapidly with $^1O_2$ and $^{\bullet}OH$ ($k = 1.07 \times 10^{10}\ M^{-1}\ s^{-1}$) but does not scavenge $O_2^{\bullet -}$ or react with H_2O_2. Rate constants for reactions involving other sunscreen active ingredients with free radicals and other oxidants have not yet been reported in the literature.

Conclusions

Sunscreen active ingredients effectively absorb and/or scatter solar ultraviolet radiation. When used properly, they can thus afford significant protection for human skin that is exposed to sunlight. Ideally, the energy acquired by absorption should be dissipated via harmless pathways such as fluorescence and internal conversion. However, based upon the limited currently available data, sunscreen active ingredients appear to undergo photofragmentation, photoisomerization, and can form reactive intermediates and undesirable stable photoproducts. There are, at present, surprisingly few published reports in the literature concerning the photochemistry and photophysics of these compounds. Much additional data from experiments conducted under carefully controlled conditions are needed in order to assess the photostability of sunscreen active ingredients that are in present use. The development of new sunscreen active ingredients that are more photochemically stable than those in present use and do not transfer absorbed energy to other molecules and form toxic species is a goal that is presently being pursued by sunscreen manufacturers.

References

1. Strange CJ. Thwarting skin cancer with sun sense. FDA Consumer 1995; 29:10-14.
2. De Gruijl FR, Van der Leun JC. Estimate of the wavelength dependency of ultraviolet carcinogenesis in humans and its relevance to the risk assessment of a stratospheric ozone depletion. Health Physics 1994; 67(4):319-325.
3. Sayre RM. Sunlight risk and how sunscreens work. Cosmetics and Toiletries 1992; 107:105-109.

4. Urbach F. Potential effects of altered solar ultraviolet radiation on human skin cancer. Photochem Photobiol 1989; 49:507-513.
5. Brown MW, Galley E. Testing UVA and UVB protection from microfine titanium dioxide. Cosmetics and Toiletries 1990; 105:69-73.
6. Drobetsky EA, Turcotte J, Châteauneuf A. A role for ultraviolet A in solar mutagenesis. Proc Natl Acad Sci USA 1995; 92:2350-2354.
7. Peak JG, Peak MJ. Comparison of initial yields of DNA-to-protein crosslinks and single-strand breaks induced in cultured human cells by far- and near-ultraviolet light, blue light and x-rays. Mutation Research 1991; 246(1):187-191.
8. Klein K. Encyclopedia of UV absorbers for sunscreen products. Cosmetics and Toiletries 1992; 107:45-65.
9. Janousek A. Sunscreens—regulatory situation worldwide. Seifen-Öle-Fette-Wachse 1991; 117(10):392-396.
10. Lowe NJ, Shaath NA. Sunscreens: Development, Evaluation and Regulatory Aspects. Cosmetic Science and Technology Series, Volume 10. New York: Marcel Dekker, 1990; 370.
11. Roberfroid M, Buc Calderon P. Free Radicals and Oxidation Phenomena in Biological Systems. New York: Marcel Dekker, 1995:65-66.
12. Mitchnick MA. Zinc oxide, an old friend to the rescue. Cosmetics and Toiletries 1992; 107:111-116.
13. Judin VPS. The lighter side of TiO_2. Chemistry in Britain 1993; 29:503-505.
14. Rudham R. Physical sunscreen materials. Presentation at the FDA Workshop on the Photochemistry and Photobiology of Sunscreens, Rockville, Maryland, USA, 19-20 September 1996.
15. Chignell CF, Kalyanaraman B, Mason RP et al. Spectroscopic studies of cutaneous photosensitizing agents-I. Spin trapping of photolysis products from sulfanilamide, 4-aminobenzoic acid and related compounds. Photochem Photobiol 1980; 32:563-571.
16. Morliere P, Avice O, Melo TSE et al. A study of the photochemical properties of some cinnamate sunscreens by steady state and laser flash photolysis. Photochem Photobiol 1982; 36:395-399.
17. Allen JM, Gossett CJ, Allen SK. Photochemical formation of singlet molecular oxygen in illuminated solutions of several commercially available sunscreen active ingredients. Chem Res Toxicol 1996; 9(3):605-609.
18. Halliwell B, Gutteridge JMC. Free Radicals in Biology and Medicine. 2nd ed. Oxford: Clarendon Press, 1995:1-20.
19. Ingold KU. Peroxy radicals. Accts Chem Res 1969; 2:1-9.
20. Sawyer DT. Oxygen Chemistry. New York: Oxford University Press, 1991:120-160.
21. Faust BC. Generation and use of simulated sunlight in photochemical studies of liquid solutions. Rev Sci Instrum 1993; 64:577-578.
22. Gasparro FP. UV-induced photoproducts of para-aminobenzoic acid. Photodermatol 1985; 2:151-157.
23. Shaw AA, Wainschel LA, Shetlar MD. The photochemistry of p-aminobenzoic acid. Photochem Photobiol 1992; 55(5):647-656.

24. Roscher NM, Lindemann MKO, Kong SB et al. Photodecomposition of several compounds commonly used as sunscreen agents. Photochem Photobiol 1994; 80:417-421.
25. Knowland J, McKenzie EA, McHugh PJ et al. Sunlight-induced mutagenicity of a common sunscreen ingredient. FEBS Letts 1993; 324(3):309-313.
26. Schallreuter KU, Wood JM, Farwell DW et al. Oxybenzone oxidation following solar irradiation of skin: photoprotection versus antioxidant inactivation. J Invest Dermatol 1996; 106:583-586.
27. Gasparro FP, Battista J. Characterization of para-aminobenzoic acid-DNA-photoadducts. Photochem Photobiol 1987; 45:49S.
28. Shaw AA, Wainschel, Shetlar MD. Photoaddition of p-aminobenzoic acid to thymine and thymidine. Photochem Photobiol 1992; 55(5):657-663.
29. Broadbent JK, Martincigh, BS, Raynor MW et al. Capillary supercritical fluid chromatography combined with atmospheric pressure chemical ionization mass spectrometry for the investigation of photoproduct formation in the sunscreen absorber 2-ethylhexyl-*p*-methoxycinnamate. Journal of Chromatography A 1996; 732:101-110.
30. Mohammad T, Baird WM, Morrison H. Photochemical covalent binding of p-methoxycinnamic acid to calf thymus DNA. Bioorganic Chem 1991; 19:88-100.
31. Gonzenbach H, Klecak G, Schwarzenbach R. Presentation at the SCS/SFC Symposium, Stratford-upon-Avon, 16-18 April 1986.
32. Schwarzenbach R, Gonzenbach H, Klecak G. New approach for testing UV-filters. Preprints of the XIVth IFSCC Congress, Barcelona, 16-19 September 1986, Vol II 1986; 845-854.
33. Selles E, Aberturas MR, Fresno MJ. Photostability of 2-ethylhexyl p-methoxycinnamate in sunscreens. An R Acad Farm 1987; 53(1):153-158.
34. Kammeyer A, Westerhof W, Bolhuis PA et al. The spectral stability of several sunscreening agents on stratum corneum sheets. Cos Sci 1987; 9(3):125-136.
35. Deflandre A, Lang G. Photoisomerization of benzylidene camphor and derivatives. Cosmetics and Toiletries 1988; 103(11):69-75.
36. Deflandre A, Lang G. Photostability assessment of sunscreens. Benzylidene camphor and dibenzoylmethane derivatives. Cos Sci 1988; 10(2):53-62.
37. Shaath NA, Fares HM, Klein K. Photodegradation of sunscreen chemicals: Solvent considerations. Cosmetics and Toiletries 1990; 105(12):41-44.
38. Turro NJ. Modern Molecular Photochemistry. Menlo Park: Benjamin/Cummings, 1978:583-593.
39. Gonzenbach H, Hill TJ, Truscott TG. The triplet energy levels of UVA and UVB sunscreens. Photochem Photobiol 1992; 16:377-379.
40. Halpern AM, Ramachandran BR. The photophysics of p-aminobenzoic acid. Photochem Photobiol 1995; 62(4):686-691.

41. Allen JM, Gossett CJ, Allen SK. Photochemical formation of singlet molecular oxygen ($^1O_2$) in illuminated aqueous solutions of *p*-aminobenzoic acid (PABA). Photochem Photobiol 1996; 32:33-37.
42. Charlier M, Hélène C. Photochemical reactions of aromatic ketones with nucleic acids and their components. I. Purine and pyrimidine bases and nucleosides. Photochem Photobiol 1972; 15(1):71-87.
43. Rutherford CE, Salter LF, Thomas RC. pH Effects on p-aminobenzoic acid photosensitized dimer formation from the free thymine base. Photochem Photobiol 1990; 52:337-343.
44. Bolton K, Martincigh BS, Salter LF. The potential carcinogenic effect of Uvinul DS49—a common UV absorber used in cosmetics. Photochem Photobiology, A: Chemistry 1992; 63:241-248.
45. Aliwell SR, Martincigh BS, Salter LF. Para-aminobenzoic acid-photosensitized dimerization of thymine I. In DNA-related model systems. Photochem Photobiol, A: Chemistry 1993; 71:137-146.
46. Aliwell SR, Martincigh BS, Salter LF. Para-aminobenzoic acid-photosensitized dimerization of thymine II. In pUC19 plasmid DNA. Photochem Photobiol, A: Chemistry 1993; 71:147-153.
47. Sutherland BM. p-Aminobenzoic acid—sunlamp sensitization of pyrimidine dimer formation and transformation in human cells. Photochem Photobiol 1982; 36:95-97.
48. Aliwell SR, Martincigh BS, Salter LF. Photoproducts formed by near-UV irradiation of thymine in the presence of p-aminobenzoic acid. Photochem Photobiol, A: Chemistry 1994; 83:223-228.
49. Aliwell SR. Para-aminobenzoic acid photosensitized dimerization of thymine. MSc thesis, University of Natal, Durban, South Africa 1991.
50. Bolton K. Studies of the photochemical reactions of thymine with selected sensitizers. MSc thesis, University of Natal, Durban, South Africa 1991.
51. Hoffmann MR, Martin ST, Choi W et al. Environmental applications of semiconductor photocatalysis. Chem Rev 1995; 95:69-96.
52. Cai R, Hashimoto K, Itoh K et al. Photokilling of malignant cells with ultrafine TiO_2 powder. Bull Chem Soc Jpn 1991; 64(4):1268-1273.
53. Cai R, Kubota Y, Shuin T et al. Induction of cytotoxicity by photoexcited TiO_2 particles. Cancer Res 1992; 52:2346-2348.
54. Anderson RR. Optics of the skin. In: Lim HW, Soter NA, eds. Clinical Photomedicine. New York: Marcel Dekker, 1993:19-31.
55. Bodannes RS, Chan PC. Singlet oxygen reacts with inhibitors of ultraviolet mediated damage to skin: p-aminobenzoic acid and its derivatives. Biochem Biophys Res Commun 1979; 87(4):1116-1123.
56. Allen JM, Egenolf S, Allen SK. Rapid reaction of singlet molecular oxygen ($^1O_2$) with p-aminobenzoic acid (PABA) in aqueous solution. Biochem Biophys Res Commun 1995; 212(3):1145-1151.
57. Hue ML, Chen YK, Chen LC et al. Para-aminobenzoic acid scavenges reactive oxygen species and protects DNA against UV and free radical damage. Nutrit Biochem 1995; 6:504-508.

Chapter 3

The Photochemical Potential of Some Sunscreens to Damage DNA

John Knowland, Peter J. McHugh and Rosemary Dunford

Introduction

It seems that the closer we look at the effects of sunlight on cells, the more we find—from immunosuppression to DNA damage. Some of the consequences, such as the generation of cyclobutane pyrimidine dimers in DNA by direct absorption of the shorter wavelengths in sunlight, are quite well understood at the molecular level; others less so. A great deal of work in this general area has used biological methods, in which test material, whether isolated cells or whole organisms, is exposed to a light source, and the results are assayed using biological criteria. For example, it is clear from such work that hairless mice develop skin tumors if they are exposed to light sources that are intended to imitate the sun. Such approaches are enormously important, but they are not the only ones available. With the recent explosion in molecular techniques for examining effects on DNA, it is becoming possible to adopt a complementary approach, in which one looks for lesions which may turn out to constitute the initial cause of the ultimate effect. In the case of human skin tumors, application of the polymerase chain reaction has made it possible to describe the mutations found in the *p53* tumor suppressor gene in the clinical specimens that have been removed from patients undergoing surgery.[1] Such information is immensely valuable because it should help to define the basic molecular cause of the tumor, and thus show what kind of DNA lesions are especially important. Building on this work, it seems reasonable to hypothesize that if we could find ways of reducing this genetic damage we could evolve strategies that would reduce our chances of developing skin tumors, whereas behavior that could increase this damage would

Sunscreen Photobiology: Molecular, Cellular and Physiological Aspects,
edited by Francis P. Gasparro.

probably increase the overall risk of skin cancer. It also seems reasonable to examine, again at the molecular level, the consequences of behavior mankind currently adopts in relation to sun exposure. In this chapter we outline the contribution which a molecular approach can make to our understanding of the field.

Background

Human skin cancers, including both the relatively curable basal cell carcinomas and squamous cell carcinomas, and the much less common but much more serious melanoma, have been increasing very rapidly throughout the world. Action to reduce a serious public health problem is urgently needed. Perhaps the simplest strategy is to avoid the sun, but many of us find this advice extremely unappealing, and are more inclined to seek the sun than avoid it. To avoid the sunburn that would otherwise ensue, we can use modern sunscreens, which are extremely efficient at eliminating sunburn because they contain materials which absorb the burning wavelengths (UVB). Consequently, a conscientious sunscreen user can confidently stay in the sun for much longer than would normally be sensible. One consequence of such behavior may be the acquisition of a much higher dose of UVA. Until recently, many sunscreens were rather poor at absorbing UVA, and this aspect must be carefully considered because it is becoming increasingly clear that UVA contributes to cancer, although the molecular mechanism is not well understood. UVA is tumorigenic to hairless mice.[2] It also induces melanoma in the melanoma-susceptible fish *Xiphophorus*, and it has been speculated from those results[3] that human users of UVB sunscreens are at risk of developing melanoma simply from the higher UVA doses that accompany longer sun exposure. Some evidence in support of this view comes from indications that users of sunbeds, which tend to emit a high proportion of UVA, appear to be at higher risk of developing melanoma.[4] Thus, sunscreens which rely principally on UVB-absorbers could yet turn out to be indirectly responsible for an increased level of some tumors.

One might hope that epidemiology would be able to shed some light on whether increased exposure to sunlight stemming from the use of sunscreens is a possibility that should be taken seriously or not, but unfortunately the evidence is very limited. Some well-controlled studies show that sunscreens reduce development of precancerous lesions in humans,[5,6] which would appear to support the view that the net effect of sunscreens is protective. However, the correlation between precancerous lesions and tumor formation is not necessarily an exact one. On the other hand, a recent epidemiological study suggested, very tentatively, that sunscreen use could constitute a risk factor for melanoma.[7] Critics of this study[8] object to the design, believing that the question posed could have introduced bias, which perhaps indicates some of the difficulties associated with an epidemiological approach. They also point out that it is essential to identify the composition of the sunscreen because, of course, one cannot assume that all sunscreens behave in exactly the same way. That is virtually impossible in an epidemiological study because people are unlikely to recall reliably the products they used many years ago. There is a very long time lag between exposure and the appearance of tumors, which means that any epidemiological study will always be subject to this limitation, making interpretation very difficult if the composition of the sunscreen is indeed a significant variable.

Another aspect of sunscreens that cannot be ignored is that they cannot destroy the energy which, by their very nature, they are intended to absorb. They have to dispose of this energy somehow, and there is increasing evidence that in so doing they may actually stimulate damage to human skin cells. For example, the UVA-absorber oxybenzone appears to inactivate the anti-oxidant enzyme thioredoxin reductase, thus undermining a natural defense mechanism.[9] For such reasons it is very important to learn as much as possible about the photochemistry of sunscreens and about what they might do to human cells, taking into account not only their ability to protect against sunburn but also their potentially harmful chemical side effects. Here, fundamental photochemistry has a great deal to contribute. Basically, when sunscreens absorb sunlight energy they may be excited to reactive species such as triplet states and free radicals, which in turn can attack many cellular components. Targets are likely to include membrane lipids as well as cellular enzymes that have already been mentioned. However, the consequences of DNA damage could be especially serious because the indications from the work outlined earlier are that such damage underlies tumor formation. Consequently, it is important to characterize the photochemistry of sunscreen chemicals in relation to their potential to attack DNA. One must of course recognize that a sunscreen which has the potential to attack DNA in some test system may not necessarily do so in normal human use, but any that do not are likely to be more attractive than those that do.

Chemical Sunscreens

The original sunscreen, *p*-aminobenzoic acid (PABA), was shown in 1982 to sensitize the formation of thymine dimers in cultured human cells.[10] It is no longer used, but a modern derivative used as a UVB filter, (Padimate-O; Fig. 3.1) is toxic and mutagenic to yeast under simulated sunlight and attacks DNA in vitro.[11] The reasons for this are clear from its structure. Chemically speaking, Padimate-O is identical to ethyl 4-dimethylaminobenzoate, a compound which generates free radicals on illumination[12] and is used as an industrial photo-initiator of polymerization for that very reason. In fact, ethyl 4-dimethylaminobenzoate was introduced as a safe photo-initiator in place of Michler's ketone (which was withdrawn as a photo-initiator when it was found to cause tumors in rats) on the grounds that it was already used in skin creams and must therefore be harmless.[13] Detailed studies on ethyl 4-dimethylaminobenzoate showed that the explanation for its photoactivity is that it contains an electron-donating group (the dimethylamino group) conjugated by the aromatic ring to an electron-withdrawing group (the carbonyl group in the ester); and it is this arrangement of functional groups that allows free radicals to form.[12] Precisely the same arrangement exists in Padimate-O, the only difference between the two compounds being that the ester group in one contains a 2-carbon unit while the ester group in the other contains a branched chain 8-carbon unit. This difference does not affect the photochemistry, but it does affect the physical properties of the compounds in a way that is highly significant from the point of view of formulating sunscreen preparations. The ethyl ester is a crystalline solid whereas Padimate-O is an oily liquid. Hence, Padimate-O is easy to incorporate into vehicles used for sunscreens, clings well to the skin and is only very slightly soluble in water. All these features make it attractive as a sunscreen, but do not alter the basic chemistry. From first principles, one would expect Padimate-O to form free radicals when illuminated. As free radicals are potentially dangerous, especially if they attack DNA, it is important to examine the effects of Padimate-O in detail, especially its potential to damage DNA. As Padimate-O penetrates skin this aspect cannot be ignored.

One of the accepted ways of asking whether compounds attack DNA is to conduct mutagenicity tests. Compounds that are mutagenic

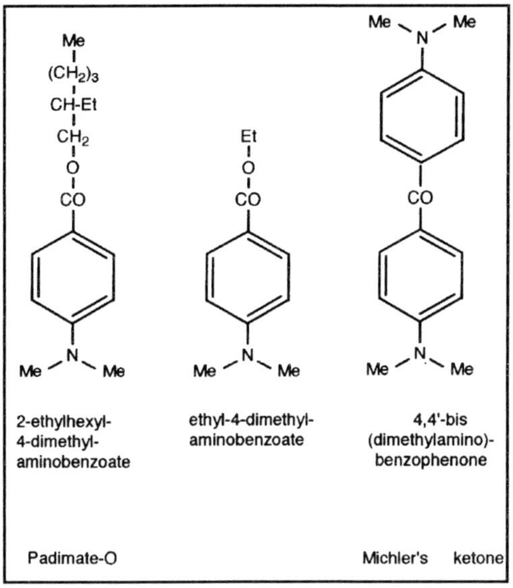

Fig. 3.1. Chemical formulae. In all three compounds, an electron-donating dimethylamino group is conjugated by an aromatic ring to an electron-withdrawing carbonyl group; a so-called Donor-Aromatic-Acceptor arrangement.[12]

to organisms such as *Salmonella* as used originally by Ames and his colleagues, or to *S. cerevisiae* as increasingly used on the grounds that it is a eukaryote, are treated with suspicion because many mutagenic compounds are also carcinogens. The correlation is by no means an absolute one, as many nonmutagens can be carcinogens,[14] but nevertheless these tests remain important. In the absence of illumination, Padimate-O was not mutagenic in normal Ames tests, and, given its photochemistry, those findings were not surprising. As soon as it is illuminated, however, it behaves very differently. It kills yeast cells in a strictly light-dependent fashion. The effects immediately disappear if the light is switched off during the experiment, and they are increased if the cells are rapidly dividing or if they lack a DNA repair pathway. Such findings suggest, but do not themselves prove, that its effects are mediated, at least in part, by attack on DNA. This suggestion is confirmed by the finding that illuminated Padimate-O induces mutations in yeast.[11]

The demonstration that Padimate-O is photomutagenic to yeast is interesting but does not explain the molecular basis of its effects. However, such information could be invaluable when it comes to probing its effects on human cells because we need to know what lesions to expect if we are to stand any chance of finding them. Does it, for example, increase the formation of thymine dimers in the way that *p*-aminobenzoic acid does, or does it induce the formation of some other kind of lesion? The simplest way to find out what

Padimate-O can do to DNA is to illuminate DNA in vitro. The knowledge obtained in this way lays the molecular foundation that is needed to explore its biological potential. Of course, we cannot be certain that when Padimate-O is illuminated in vivo it will, chemically speaking, behave in precisely the same way as it does in vitro, but without any understanding of the fundamental chemistry we are unlikely to make much progress. Hence the importance of the in vitro, chemical approach. We expect illuminated Padimate-O to generate free radicals and indeed it does. The evidence for this includes direct examination by electron spin resonance spectroscopy[15] and suppression of damage to DNA by known free radical quenchers (Fig. 3.2). As expected for a producer of free radicals, it does not increase the formation of cyclobutane dimers or [6-4] photoproducts. These findings are particularly interesting because the majority of work on photodamage to DNA has focused on thymine dimers and [6-4] photoproducts, which are formed as a result of direct absorption of light. Consequently, work based purely on assays for these lesions will not reveal the molecular effects of Padimate-O. It is even possible that Padimate-O could, by absorbing UVB, protect against the formation of photoproducts which arise from the direct absorption of light, such as thymine dimers, but simultaneously catalyze the formation of other lesions. What are the effects of Padimate-O? It generates direct strand breaks and other lesions that can be revealed by cleaving at the lesions, but does not sensitize the formation of thymine dimers to any detectable extent.[16] Cleaving specifically at these lesions is not particularly easy because they are comparatively rare and demand very precise methods. It is important to use methods that maximize specific cleavage and minimize nonspecific cleavage. The conventional cleavage reagent is 1 M piperidine at 90°C, and it can often be used with success, particularly when the number of lesions is relatively high, which is why, of course, it is so successful in Maxam-Gilbert DNA sequencing. However, it is not very satisfactory for detecting rare lesions in DNA because of significant nonspecific cleavage caused by the high pH which it generates. Hydroxyl ions alone can cleave undamaged DNA because they can damage or displace some bases, and indeed alkali was used in some of the original Maxam-Gilbert DNA sequencing methods for that very reason.[17] Therefore, it is important to find alternative methods for cleaving at lesions which do not run the risk of nonspecific cleavage. We have looked at a variety of reagents and have found one in particular,

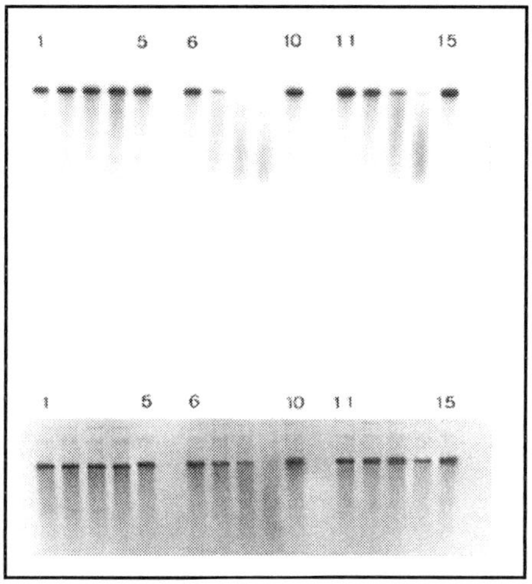

Fig. 3.2. Damage to end-labeled DNA. End-labeled, double-stranded M13 mp18 DNA was illuminated as described[11] and analyzed by electrophoresis on alkaline agarose gels. Top panel. Lanes 1-5; simulated sunlight alone for 0, 20, 40 and 60 minutes. Lanes 6-10; simulated sunlight plus 50 mM benzophenone used as a positive control (benzophenone is known to induce strand breaks in DNA). Lanes 11-15; simulated sunlight plus 50 mM Padimate-O. Bottom panel. Treatments were as described for the top panel except that the quenching agent ethanol[21] was present at 2% v/v in all cases.

N,N'-dimethylethylenediamine (DMED), which cleaves lesions in DNA at neutral pH, and does not cleave nonspecifically. At neutral pH, one of its two amino-groups is positively charged, facilitating binding to DNA, and the other remains uncharged, which means that it can catalyze elimination reactions. We have examined the reaction in detail and find that the pattern of elimination can, to some extent, be adjusted at will.[18] For example, DMED cleaves abasic sites by β-elimination at 37°C and by β,δ-elimination at 90°C (Fig. 3.3). A certain amount of cleavage also proceeds by 4',5'-cyclization, when the fragments released 3' to the lesion bear cyclic deoxyribose phosphates at their 5'-ends. Although this is a relatively minor pathway, its existence does mean that the simplest autoradiographic cleavage patterns are obtained when the DNA under examination is labeled at the 5' rather than the 3'-end. It also means that if further analysis by ligation-mediated PCR[19] is contemplated, when the 3'-fragments released are ligated to another molecule, some fragments may be underestimated because if they are blocked by cyclic phosphates they will not ligate.

However, these are not serious drawbacks, and DMED has proved to be a useful addition to the reagents available for studying lesions in DNA. It cleaves at photoproducts induced by ultraviolet light alone, and also at lesions induced by free radicals from Padimate-O and simulated sunlight.[16] Precisely what those lesions are is still not clear, and indeed, it is likely that there is a family of

Fragments released by elimination or cyclization at an abasic site

B1 and B2 represent normal bases, dR an abasic site, dR an unsaturated aldehyde.

P represents free phosphate; Δ indicates a cyclic phosphate.

```
                         cleavage points
              4',5'        δ          β        3',4'
5'------P--dR(B1)----|----P----|----dR----|----P----|---dR(B2)--P------3'
```

β-elimination yields:	P--dR(B1)--P--dR			P--dR(B2)--P
β,δ-elimination yields:	P--dR(B1)--P	dR		P--dR(B2)--P
3',4' cyclization yields:	P--dR(B1)--P--dRΔP			HO--dR(B2)--P
which may be followed by elimination to give:	P--dR(B1)--P--dR		P	HO--dR(B2)--P
which may be followed by elimination to give:	P--dR(B1)--P	dR	P	HO--dR(B2)--P
4',5' cyclization yields:	P--dR(B1)--OH			PΔdR--P--dR(B2)--P
which may be followed by elimination to give:	P--dR(B1)--OH	P--dR		P--dR(B2)--P

At abasic sites, **β-elimination** is favored at 37°C, and **β,δ-elimination** is favored at 90°C. The cyclization reactions leave cyclic phosphates on the upstream and downstream fragments as indicated. Cyclic phosphates are not convenient substrates for further molecular manipulations. For details see ref. 18.

Fig. 3.3. The types of cleavage catalyzed by N,N'-dimethylethylenediamine at abasic sites.

lesions, because work on the damage inflicted by hydroxyl radicals on DNA shows that there are many possible products.[20] Nevertheless, this chemical work helps to explain, at the molecular level, the results found using illuminated Padimate-O and yeast, and will act as a guide for further studies designed to probe the effects of Padimate-O and other sunscreens on DNA in human cells. One potentially very important point also emerges from this work. It is that DMSO quenches, very efficiently, the effects of Padimate-O, the reason being that DMSO is a very efficient scavenger of hydroxyl radicals.[21] This could have major implications for photomutagenicity tests, because they are often conducted in the presence of DMSO. In a common version of Ames tests, the test chemical is added to auxotrophic cells embedded in a thin layer of agarose spread over a thicker layer which does not contain cells but does contain materials needed for growth.[22] The mutagenicity of the chemical is then scored by counting the revertant colonies that appear. As many materials are not very water-soluble, Ames and his colleagues examined a number of alternative solvents, searching for nontoxic ones which would dissolve test chemicals efficiently but would not interfere with the tests. One of the best was DMSO.[23] As the mutagenicity tests in which DMSO was used were not designed to detect short-lived mutagens such as those generated by illumination, the ability of DMSO to quench some excited states was of no relevance. However, it becomes highly relevant in photomutagenicity tests, and it is important to ask exactly how they are conducted. Usually, illumination is introduced as a simple addition to the overall procedure. The plates are illuminated after the thin layer of agarose, containing 2-3% DMSO by volume, has been spread over the bottom layer.[24] Depending on the delay between preparing the plates and illuminating them, and on the rate at which DMSO diffuses out of the top agarose and into the bottom agarose, the cells may well be illuminated in the presence of a concentration of DMSO which is more than sufficient to quench any hydroxyl radicals that may be formed. If illumination is conducted immediately after preparing the plates, the DMSO concentration will be 2-3%. Hence, if a sunscreen generates free radicals when it is illuminated conventional mutagenicity tests may produce negative, but false, results. It should also be remembered that another quencher of free radicals is glucose, a standard component of the bottom agarose, which again could complicate the interpretation of the results. These considerations underscore the importance of understanding the photochemistry of sunscreen compounds if the objective is to study their potential to inflict damage on DNA. It may not be appropriate to simply introduce an illumination step into a normal Ames test. One might also add that there is another potential pitfall of DNA damage assays which this time applies to in vitro studies. The conventional buffer that most workers would automatically use for such work is Tris. However, Tris buffers can also quench free radicals,[21] and indeed, completely suppress the effects of Padimate-O on DNA,[16] so that it is essential to avoid Tris if meaningful results are to be obtained from in vitro studies.

'Physical' Sunscreens

Concern about the possible adverse effects of organic chemicals in sunscreens has encouraged many manufacturers to rely more and more on inorganic alternatives, notably zinc oxide and titanium dioxide. They are perceived as harmless pigments which cannot enter skin and which cannot be affected by light energy in the way that organic chemicals may be. Zinc oxide has been used for years as an ingredient of many skin ointments and titanium dioxide is a standard ingredient of many paints, attractive because of its great opacity and low price. Both of these, and especially titanium dioxide, are now widely used in sunscreens, and are sometimes marketed as "nonchemical materials", implying a high degree of safety. However, both compounds are semiconductors which can absorb light and generate reactive species from its energy. Their interaction with absorbed light is related to the "band gap". The band gap is measure of the minimum energy in electron volts required to promote a single electron from the valence band to the conduction band, and for both zinc oxide and titanium dioxide is about 3 eV. As a band gap of 1 eV corresponds to a wavelength of 8.06616×10^3 cm^{-1}, compounds with band gaps in the region of 3 eV are excitable by light at wavelengths below about 380 nm. In sunlight at sea level, there is negligible energy below about 295 nm, but it increases rapidly between 295 nm and 400 nm, so that both zinc oxide and titanium dioxide are susceptible to excitation by UVB and UVA in natural sunlight.

The precise value of the band gap for titanium dioxide, and hence the excitability, depends upon the crystal structure, being 3.23 electron volts for anatase and 3.06 eV for rutile,[25] which correspond to wavelengths of 379 and 400 nm, respectively. When light at or below the critical wavelength promotes a single electron from the valence band to the conduction band it leaves a positively charged space, or hole, behind. Usually, the electron recombines with the hole, but sometimes the hole migrates to the surface of the particle, where it can react with absorbed species. In an aqueous environment it can react with water or hydroxyl ions, forming hydroxyl radicals.[26] For example:

$$\text{hole}^+ + OH^- \longrightarrow \text{hole} + OH^\bullet$$

Such processes are well known for titanium dioxide exposed to both artificial UV light and natural sunlight and indeed, the photocatalytic potential of titanium dioxide has been widely studied and exploited because it can be very high. Illumination of titanium dioxide suspensions with sunlight can degrade organic materials and purify drinking water,[27] while illumination with short-wave UV kills human cells.[28] This work shows that the distinction between organic, 'chemical' sunscreens and inorganic, 'physical' sunscreens, attractive though it may be to those who market them, is not based on any significant difference. Both varieties have the potential to produce reactive species that can attack biological materials when they are exposed to normal sunlight.

An otherwise comprehensive evaluation of the safety of titanium dioxide concluded that it is not mutagenic and hence cannot damage DNA, but it did not examine the effects of sunlight.[29] In view of its behavior in sunlight, it is essential to do so, taking into account the physical form in which it is used. In sunscreens, titanium dioxide is usually used as 'micronized' particles in the size range 20-50 microns for very good cosmetic reasons. Such particles are small compared with the wavelength of the incident light and so they scatter it according to Rayleigh's law, which states that the intensity of the scattered light is inversely proportional to the fourth power of the wavelength. Consequently, micronized particles do not scatter visible light and are virtually invisible on the skin. However, they still absorb UVB and UVA, and absorption can be followed by the generation of hydroxyl radicals. Whether this happens or not depends upon a number of factors, many of which are imperfectly understood. For example, rutile titanium dioxide is much less photoactive than anatase,[30] despite its similar band gap, and appears to dissipate much of the energy which it absorbs by means which do not involve the formation of reactive holes. Sunscreen titanium dioxide preparations are often coated with other materials (oxides of aluminium, silicon or zirconium) which can reduce photoactivity. Thus, it may be possible to eliminate the photochemical potential of titanium dioxide, or at least to reduce it to acceptable levels. These possibilities need to be systematically explored, bearing in mind, of course, that the levels of DNA damage that could be harmful for human cells are extremely small. We have started to examine these aspects. We find that titanium dioxide prepared by adding titanium (IV) chloride to water (principally anatase) attacks DNA very efficiently when illuminated, as one would expect. However, we also find that titanium dioxide found in a commercially available sunscreen can be equally damaging (Fig. 3.4). That does not mean, of course, that all titanium dioxide preparations will necessarily behave like this, but it does mean that there is a need for precise chemical research in this area.

Another very important question here is whether micronized particles of titanium dioxide and zinc oxide can penetrate human

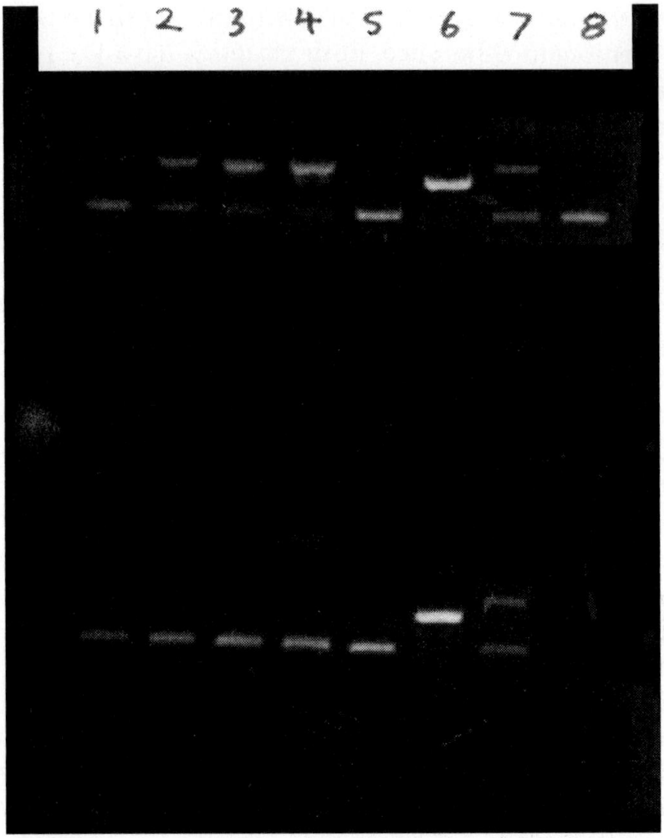

Fig. 3.4. Damage to a supercoiled plasmid catalyzed by titanium dioxide from a commercial sunscreen. The supercoiled plasmid pBluescript II SK + (Stratagene) was illuminated as in Figure 3.2 in the presence of titanium dioxide extracted from a commercial sunscreen by washing it with a 1:1 mixture of chloroform and ethanol, followed by ethanol, water and drying. The concentration during illumination was 0.0125% w/v. Samples were analyzed by electrophoresis on neutral agarose gels in the presence of ethidium bromide. Single-strand nicks in the DNA result in conversion of the plasmid from a supercoiled form to a relaxed form with lower electrophoretic mobility. Top panel: Lanes 1-4; samples taken after 0, 20, 40 and 60 minutes illumination with titanium dioxide. Lanes 5-7; supercoiled, linear and relaxed standards. There is some supercoiled as well as relaxed DNA in lane 7. Lane 8; plasmid incubated with titanium dioxide in the dark for 60 minutes. Bottom panel: Lanes 1-4; samples taken after 0, 20, 40, 60 minutes illumination without titanium dioxide. Lanes 5-7; supercoiled, linear and relaxed standards.

skin. The natural assumption is that fine solids, even micronized ones which are smaller than many viruses, cannot, but there is little published evidence to support this belief. Presumably, solubility must play a part, but even though titanium dioxide is virtually insoluble in water there are suggestions from biopsies of clinical material followed by energy dispersive X-ray analysis or electron probe microanalysis that it can pass through skin.[31,32] The findings are far from conclusive and the possible route of entry is not at all clear, but it may not be safe to assume that titanium dioxide cannot enter skin. Zinc oxide has a small but finite solubility in water, and there is some evidence from direct analysis of zinc content [33] and from autoradiographic studies using ^{65}ZnO that it may be able to pass through rabbit skin into the dermis and below, probably through hair follicles.[34] This whole question of permeability of human skin to inorganic sunscreens needs to be examined carefully, with methods that are sensitive enough to detect the very low levels that are relevant to DNA damage. What is established is that particles of titanium dioxide as large as 220 nm can enter human cells in culture,[35] and so it seems entirely plausible that if titanium dioxide does pass through skin it could enter cells under the skin.

Conclusions

The main conclusion to emerge from this brief survey of the chemical potential of illuminated sunscreens, both organic and inorganic, is that we know remarkably little about them, especially considering how widely they are used. But there is sufficient evidence from studies on their potential to attack DNA to warrant more detailed work than has been possible until recently. The belief that they are safe to apply to human skin and protect against the effects of the sun, whether in respect to oxidative damage to the cytoplasm or in respect to potentially more sinister damage to the nucleus, is founded mainly on experiments in animals and microorganisms. It is now time to apply the currently available molecular methods to human cells.

Acknowledgments

We thank Mrs. G. Harbour for technical assistance and the Wellcome Trust and the Medical Research Council for financial support.

References

1. Brash DE et al. A role for sunlight in skin cancer: UV-induced p53 mutations in squamous cell carcinoma. Proc Natl Acad Sci USA 1991; 88:10124-10128.
2. Sterenborg HJCM, van der Leun JC. Tumorigenesis by a long wavelength UVA source. Photochem Photobiol 1990; 51:325-330.

3. Setlow RB, Woodhead AD. Temporal changes in the incidence of malignant melanoma: explanation from action spectra. Mutat Res 1994; 307:365-374.
4. Bajdik CD, Gallagher RP, Astrakianakis G, Hill GB, Fincham S, McLean DI. Nonsolar ultraviolet radiation and the risk of basal and squamous cell skin cancer. Br J Cancer 1996; 73:1612-1624.
5. Thompson SC, Jolley D, Marks R. Reduction of solar keratoses by regular sunscreen use. N Engl J Med 1993; 329:1147-51.
6. Naylor MF et al. High sun protection factor sunscreens in the suppression of actinic neoplasia. Arch Dermatol 1995; 131:170-5.
7. Westerdahl J, Olsson H, Msbck A, Ingvar C, Jonsson N. Is the use of sunscreens a risk factor for malignant melanoma? Melanoma Res 1995; 5:59-65.
8. Roberts LK, Stanfield JW. Suggestion that sunscreen use is a melanoma risk factor is based on inconclusive evidence. Melanoma Res 1995; 5:377-379.
9. Sundaram C, Koster W, Schallreuter KU. The effect of UV radiation and sun blockers on free radical defence in human and guinea pig epidermis. Arch Dermatol Res 1990; 282:526-31.
10. Sutherland BM. p-aminobenzoic acid-sunlamp sensitization of pyrimidine dimer formation and transformation in human cells. Photochem Photobiol 1982; 36:95-97.
11. Knowland J, McKenzie EA, McHugh PJ, Cridland NA. Sunlight-induced mutagenicity of a common sunscreen ingredient. FEBS Lett 1993; 324:309-313.
12. Forster M, Hester RE. Photoionization of dimethylaniline and of the donor-aromatic-acceptor molecules p-(CH3)2NC6H1R, R=CN, COOEt, CHO, NO2, in glasses of ethanol and ethanediol/water at 77-125 K. J Chem Soc, Faraday Trans. 1981; (2)77:1521-1534.
13. Phillips R. Photopolymerization. J Photochem 1984; 25:79-82.
14. Zeiger E. Carcinogenicity of mutagens: predictive capability of the Salmonella mutagenesis assay for rodent carcinogenicity. Cancer Res 1987; 47:1287-1296.
15. McHugh PJ. D. Phil. Thesis, Oxford, 1996.
16. McHugh PJ, Knowland J. Characterisation of DNA damage inflicted by free radicals from a mutagenic sunscreen ingredient and its location using an in vitro reversism assay. Photochem Photobiol.
17. Maxam AM, Gilbert W. A new method for sequencing DNA. Proc Natl Acad Sci USA 1977; 74:560-564.
18. McHugh PJ, Knowland J. Novel reagents for chemical cleavage at abasic sites and UV photoproducts in DNA. Nuc Acid Res 1995; 23:1664-1670.
19. Pfeifer GP. Technologies for Detection of DNA Damage and Mutations. New York: Plenum Press, 1996.
20. Hutchinson F. Chemical changes induced in DNA by ionizing radiation. Prog Nucleic Acids Res and Mol Biol 1985; 32:115-154.

21. Halliwell B, Gutteridege JMC. Free Radicals in Biology and Medicine. Oxford: Clarendon Press, 1989.
22. Maron DM, Ames BN. Revised methods for the Salmonella mutagenicity test. Mut Res 1983; 113:173-215.
23. Maron D, Katzenellenbogen J, Ames BN. Compatibility of organic solvents with the Salmonella/microsome test. Mut Res 1981; 88:343-350.
24. Dean SW, Lane M, Dunmore RH, Ruddock SP, Martin CN, Kirkland DJ, Loprieno N. Development of assays for the detection of photomutagenicity of chemicals during exposure to UV light -1. Assay development. Mutagenesis 1991; 6:335-41.
25. Strehlow WH, Cook EL. Compilation of energy band gaps in elemental and binary compound semiconductors and insulators. J Phys Chem Ref Data 1973; 2:163-199.
26. Brezova V, Stasko A. Spin trap study of hydroxyl radicals formed in the photocatalytic system TiO2-water-p-cresol-oxygen. J Catalysis 1994; 147:156-162.
27. Hoffman MR, Martin ST, Choi W, Bahnemann DW. Environmental applications of semiconductor photocatalysis. Chem Rev 1995; 95:69-99.
28. Cai R, Kubota Y, Shuin T, Sakai H, Hashimoto K, Fujishima A. Induction of cytotoxicity by photoexcited TiO_2 particles. Cancer Res 1992; 52:2346-2348.
29. Titanium dioxide. IARC Monogr Eval Carcinog Risks Humans 1989; 47:307-326.
30. Sclafani A, Palmisano L, Davi E. Photocatalytic degradation of phenol by TiO_2 aqueous dispersions: rutile and anatase reactivity. New J Chem 1990; 14:265-268.
31. Moran CA, Mullick FG, Ishak KG, Johnson FB, Hummer WB. Identification of titanium in human tissues: probable role in pathologic processes. Hum Pathol 1991; 22:450-454.
32. Dupre A, Touron P, Daste J, Lassere J, Bonafe JL, Viraben R. Titanium pigmentation. An electron probe microanalysis study. Arch Dermatol 1985; 121:656-658.
33. Agren MS. Percutaneous absorption of zinc from zinc oxide applied topically to intact skin in man. Dermatologica 1995; 180:36-39.
34. Kapur SP, Bhussry BR, Rao S, Harmouth-Hoene E. Percutaneous absorption of zinc in rabbit skin. Proc Soc Exp Biol Med 1974; 145:932-937.
35. Cai R, Hashimoto K, Itoh K, Kobota Y, Fujishima A. Photokilling of malignant cells with ultrafine TiO_2 powder. Bull Chem Soc Japan 1974; 64:1268-1273.

CHAPTER 4

Sunscreens and the Prevention of Erythema

Cheryl F. Rosen

Sunscreens are defined by their ability to prevent ultraviolet radiation-(UVR) induced erythema, or sunburn. It is this characteristic that is essential to labeling a compound a sunscreen. The ability of sunscreens to reduce sunburn is well established, confirmed by "extensive human experience."[1]

Much work has been devoted to determining whether the use of a sunscreen results in further benefit beyond the prevention of erythema, such as the prevention of skin cancer, actinic keratoses, wrinkling and tanning.

The Sun Protection Factor (SPF) grading refers to protection against erythema, primarily UVB-induced, as UVB is 1000 times more erythemogenic than UVA. The SPF is the ratio of the amount of ultraviolet radiation to produce a minimal erythema response, 24 hours after exposure, in an area of sunscreen-protected skin to the amount required to produce the same degree of erythema in unprotected skin. A standardized procedure, with a defined radiation source with a stated spectral power distribution, defined skin types of the human volunteers, and a specified amount of sunscreen (most often 2 $\mu l/cm^2$) has not been universally agreed upon, but these variables must be outlined in studies of sunscreen SPF values. An artificial source of UVR, a solar simulator which produces a range of wavelengths which approximates solar ultraviolet spectral output, is preferred over natural sunlight in the U.S. FDA. recommendations[2] where an attempt was made to standardize the procedure. A sunscreen with an SPF of 15 prevents 92% of the UV from penetrating the skin.

Sunscreen Photobiology: Molecular, Cellular and Physiological Aspects,
edited by Francis P. Gasparro.

SPFs obtained through outdoor solar testing are generally lower than indoor solar-simulator values, and of course the SPF obtained during actual use may also differ from that on the label, primarily due to the amount and frequency of application of the sunscreen to the skin.[4] The ability of a sunscreen to deliver its advertised SPF may be affected by the thickness of application by the user. The amount of sunscreen applied in a study using 50 healthy volunteers found that the use of product ad libitum resulted in SPF values which were only 50% of those achieved with a standard application of 2 mg/cm^2.[5]

Because the SPF value is primarily indicative of protection against UVB, a variety of testing methods have been developed in order to measure the effectiveness of sunscreen products to protect skin from the effects of UVA. As the lower erythemogenic ability of UVA leads to practical problems such as length of exposure time required, alternatives to using UVA-induced erythema as the end point have been developed. There is not one standard procedure that is followed to give a standardized UVA Protection Factor. Different protocols have chosen a variety of endpoints, including erythema, immediate pigment darkening, delayed tanning, inhibition of ^3H thymidine incorporation to evaluate sunscreen efficacy in blocking UVA effects.[6-8] In some studies the skin is pretreated with UVA photosensitizers (e.g. psoralen, anthracene) to lower the threshold of UVA to produce erythema.[9] The photosensitizer is given either orally or topically. It is problematic to use photosensitizers, as an exaggerated protection factor may be obtained when the absorption spectrum of the sunscreen coincides with the action spectrum of the sensitizer.[10] In other studies, nonsensitized erythema has been used as the endpoint in an animal model and in human volunteers.[6,7] Photoprotection against UVA-induced erythema and pigmentation was compared in photosensitized skin and nonsensitized skin in one study.[11] Different results were obtained, with higher UVA protection factors noted in photosensitized skin. Parsol 1789 and Eusolex 8020, another dibenzoylmethane, provided better protection than benzophenone in photosensitized skin. However, in nonsensitized skin all three compounds protected equally well against erythema, with Parsol 1789 and Eusolex 8020 being slightly better against immediate and delayed pigmentation.[11] When immediate pigment darkening was used as the endpoint, Parsol 1789 plus benzophenone-3 was found to be more effective than benzophenone-3 alone.[8] A recent study used a population of photosensitive patients with erythropoietic protoporphyria and chronic actinic dermatitis, to evaluate sunscreen efficacy against UVB, UVA and visible light.[12]

In a variety of studies, prevention of erythema by a sunscreen product was determined as a preliminary step, where the study was more particularly interested in the prevention of another UV-inducible alteration in the skin.[13,14] These studies provide further information on the prevention of erythema by a variety of compounds. For example, in a study to examine the ability of sunscreens to prevent UV-induced herpes labialis recurrences, patients did not develop erythema at all within sunscreen-treated areas. The sunscreen used was SPF 15, containing glyceryl p-aminobenzoate, padimate O and oxybenzone.[13] In a recent study primarily focused on the prevention of UVB-induced immunosuppression, human subjects exposed to three MEDs from a bank of fluorescent bulbs emitting >70% between 290 and 320 nm, not surprisingly, did not develop erythema at any irradiated sites, using a sunscreen with SPF 29, containing octyl methoxycinnamate, oxybenzone and octyl salicylate.[14] In a study of human skin examining sunburn cell formation, using sunscreens containing either 8% homomethylsalicylate, 5% PABA or 10% sulisobenzone, erythema was prevented in all 10 volunteers, exposed to 3 minimal erythema doses (MED) of UVA and UVB emitted by a xenon arc solar simulator.[3]

Prevention of erythema, of course, does not equal prevention of UV-induced changes in the skin. Alteration in the activity of epidermal enzymes, such as glucose-6-phosphate dehydrogenase, and thickening of the stratum corneum occurred in spite of prevention of erythema.[15] This would imply that the wavelengths leading to the induction of erythema are absorbed by the sunscreening compounds, while non-absorbed wavelengths are involved in inducing other effects.

In many studies where the effects of sunscreens on a variety of ultraviolet-induced alterations in murine skin are examined, edema is used as an endpoint instead of a visual assessment of erythema.[16,17] In these studies, sunscreens were able to consistently prevent the induction of edema by ultraviolet radiation. The sunscreens studied consisted of O-PABA, 2-EHMC and a benzophenone preparation. The source of UV was a bank of FS40 sunlamps with a peak emission at 313 nm.[16,17]

In a study by Reeve et al, both erythema and edema were observed. Edema was quantified, measured as skinfold thickness.[18,19] With both sunscreens tested (2-EHMC and o-PABA), no erythema was detected after three consecutive daily exposures to UVB from FL40SE fluorescent tubes.[18]

In other work, mice were exposed to either a xenon arc solar simulator or fluorescent UVB source, and minimal erythema doses were determined. Two commercial sunscreen preparations containing Padimate O and oxybenzone were able to provide protection of SPF 5 or 13,[20] similar to their labeled values.

In many studies in both mice and humans where endpoints other than erythema are used (such as p53 and p21 expression), erythema or edema due to UV alone, and not due to a contact sensitizer, are not measured or reported.[21,22] Sunscreens are considered to provide "universal protection" from UV-induced erythema and edema."[19]

Other factors besides the actual UV absorbing, scattering and reflecting compounds influence the effectiveness of sunscreens.[23,24] The time between application of the sunscreen and solar exposure may be important. Some authors have recommended at least a 20 minute period between application of the sunscreen compound and sunlight exposure.[24] The composition of the vehicle and the substantivity of a product, that is the ability to remain adsorbed on the skin in the face of water exposure, are very important. The thickness of the applied sunscreen is standardized in SPF testing to 2 $\mu l/cm^2$, but in actual usage there can be great variability in the thickness applied. The test subjects used can vary in their constitutive amount of melanin and in their ability to tan. The radiation source greatly affects the SPF values.

In summary, sunscreens are effective in preventing erythema, if used correctly. However, the ability to prevent erythema may lead to people increasing their exposure time because they are not burning. Therefore, it is important to educate the consumer to realize that protection of erythema does not equal protection against all the deleterious effects of sunlight exposure.

References

1. International Agency for Research on Cancer World Health Organization: Solar and ultraviolet radiation. IARC Monographs on the Evaluation of Carcinogenic Risks to Humans 1992; (1)55:285-290.
2. Lowe NJ. Photoprotection. Seminars Dermatol 1990; 9:78-83.
3. Grove GL, Kaidbey KH. Sunscreens prevent sunburn cell formation in human skin. J Invest Dermatol 1980; 75:363-364.
4. Kaplan LA. Suntan, sunburn, and sun protection. J Wild Med 1992; 3:173-196.

5. Stenberg C, Larko O. Sunscreen application and its importance for the Sun Protection Factor. Arch Dermatol 1985; 121:1400-1402.
6. Chew S, DeLeo VA , Harber LC. An animal model for evaluation of topical photoprotection against ultraviolet A (320-380 nm) radiation. J Invest Dermatol 1987; 89:410-414.
7. Cole C, Van Fossen R. Measurement of sunscreen UVA protection: an unsensitized human model. J Amer Acad Dermatol 1992; 26:178-184.
8. Kaidbey KH, Barnes A. Determination of UVA protection factors by means of immediate pigment darkening in normal skin. J Amer Acad Dermatol 1991; 25:262-266.
9. Jarratt M, Hill M, Smiles K. Topical protection against long-wave ultraviolet A. J Amer Acad Dermatol 1983; 9:354-360.
10. Urbach F, Cole CA. Effects of light source and photosensitizer on predicted protection factors of UVA sunscreens. Photochem Photobiol 1986; 43(S):86S.
11. Kaidbey K, Gange RW. Comparison of methods for assessing photoprotection against ultraviolet A in vivo. J Amer Acad Dermatol 1987;16:346-353.
12. Diffey BL, Farr PM. Sunscreen protection against UVB, UVA and blue light: an in vivo and in vitro comparison. Brit J Dermatol 1991; 124:258-263.
13. Rooney JF, Bryson Y, Mannix ML et al. Prevention of ultraviolet-light-induced herpes labialis by sunscreen. Lancet 1991; 338:1419-1422.
14. Whitmore SE, Morison WL. Prevention of UVB-induced immunosuppression in humans by a high Sun Protection Factor sunscreen. Arch Dermatol 1995; 131:1128-1133.
15. Pearse AD, Marks R. Response of human skin to ultraviolet radiation: dissociation of erythema and metabolic changes following sunscreen protection. J Invest Dermatol 1983; 80:191-194.
16. Wolf P, Yarosh DB, Kripke ML. Effects of sunscreens and a DNA excision repair enzyme on ultraviolet radiation-induced inflammation, immune suppression, and cyclobutane pyrimidine dimer formation in mice. J Invest Dermatol 1993; 101:523-527.
17. Wolf P, Donawho CK, Kripke ML. Analysis of the protective effect of different sunscreens on ultraviolet radiation-induced local and systemic suppression of contact hypersensitivity and inflammatory responses in mice. J Invest Dermatol 1993; 100:254-259.
18. Reeve VE, Bosnic M, Boehm-Wilcox C, Ley RD. Differential protection by two sunscreens from UV radiation-induced immunosuppression. J Invest Dermatol 1991; 97:624-628.
19. Reeve VE, Boehm-Wilcox C, Bosnic M, Reilly WG. Differential photoimmunoprotection by sunscreen ingredients is unrelated to epidermal cis urocanic acid formation in hairless mice. J Invest Dermatol 1994; 103:801-806.
20. Fisher MS, Menter JM, Willis I. Ultraviolet radiation-induced suppression of contact hypersensitivity in relation to padimate O and oxybenzone. J Invest Dermatol 1989; 92:337-341.

21. Ponten F, Berne B, Ren Z-P, Nister M, Ponten J. Ultraviolet light induces expression of p53 and p21 in human skin: efect of sunscreen and constitutive p21 expression in skin appendages. J Invest Dermatol 1995; 105:402-406.
22. Gurish MF, Roberts LK, Krueger GG, Daynes RA. The effect of various sunscreen agents on skin damage and the induction of tumor susceptibility in mice subjected to ultraviolet irradiation. J Invest Dermatol 1981; 76:246-251.
23. Pathak MA. Sunscreens: Topical and systemic approaches for protection of human skin against harmful effects of solar radiation. J Amer Acad Dermatol 1982; 7:285-312.
24. Taylor CR, Stern RS, Leyden JJ, Gilchrest BA. Photoaging/photodamage and photoprotection. J Amer Acad Dermatol 1990; 22:1-15.

CHAPTER 5

Do Broad Spectrum Sunscreens Allow Ultraviolet-Induced Photodamage in the Absence of Erythema?

Lorraine H. Kligman

Introduction

Broad spectrum, high sun protection factor (SPF) sunscreens have been shown, unequivocally, to prevent sunburn as long as the UV dose does not exceed the protective factor of the sunscreen.[1,2] Although the acute effects of sunlight can be readily studied in humans, assessment of sunscreen efficacy against the damage caused by chronic irradiation requires an animal model. The hairless mouse has been and continues to be a useful model for these purposes.[3] High SPF products have been shown to prevent UVB-induced carcinogenesis in these mice.[4,5] The model has been established as a relevant one for the study of photoaging[3,6,7] because the dermal connective tissue changes closely resemble those found in human sun worshippers.[8] Sunscreen prevention of UVB-induced photoaging has been repeatedly demonstrated.[6,7,9] Dermal damage induced by long-term UVA exposure has also been prevented by sunscreens containing UVA-absorbing molecules.[10,11] Similarly, sunscreen protection against chronic exposure to solar simulating radiation (UVB plus UVA) has been shown.[12] In summary, experimental studies with the hairless mouse have reported that sunscreens provide virtually complete protection against photodamage resulting from chronic exposure to UV radiation. However, a retrospective examination of such studies makes it clear that a number of issues need to be explored before such a conclusion can be made with confidence.

Sunscreen Photobiology: Molecular, Cellular and Physiological Aspects, edited by Francis P. Gasparro.

What Are the Issues?

One of the prime issues in the earlier studies is the use of UV doses that are far below the protection limit of the sunscreens. In one study, a UVB dose of less than 1 minimal erythema dose (MED), as determined in unprotected animals, was used to irradiate mice protected with an SPF 15 sunscreen for 33 weeks.[7] The sunscreen blocked the physical changes measured, which included increases in skinfold thickness and transepidermal water loss as well as skin wrinkling. Histological changes were minimal. Another study, in which an SPF 8 sunscreen was tested for up to 12 weeks with a UVB dose of 1/2 MED given five times a week, reported no significant changes in collagen metabolism compared to unirradiated controls.[9] A third study used a higher UVB dose (6 MED thrice weekly for 30 weeks) which was still considerably below the protection limit of the SPF 15 sunscreen. Although a mild epidermal hyperplasia was noted, connective tissue histologic changes produced in unprotected UV irradiated mice were prevented.

The use of UV doses far below the SPF of the sunscreen is not restricted to hairless mouse studies. Protection against UVB-induced cyclobutyl thymine dimers in the DNA of human epidermis was demonstrated with 1 MED and an SPF 10 sunscreen.[13] Similarly, protection against UVB-induced immunosuppression has been reported in humans with 3 MEDs and an SPF 29 sunscreens[14] and in haired mice with ~3 minimal edema doses (requires slightly less irradiation than an MED) and SPF 15 and 30 sunscreens.[15] The point to be made is that while sunscreen protection studies must not use UV doses that exceed the protection limit of the products, neither should very low doses be used. It is crucial to have information on how well sunscreens can be expected to perform when the user is exposed to chronic UV radiation for more than one or two hours per exposure.

Another important issue is the overwhelming use of UVB in hairless mouse studies that assess protection against photoaging by broad spectrum sunscreens.[3,6,7,9,16] UVB sources (290-320 nm) are an economical, effective and useful tool with which to study UV-induced changes in skin. They are, however, clearly not representative of the solar radiation that impinges on human skin, as they lack the large UVA component (320-400 nm) present in sunlight. It is well known that UVA has a significant impact on skin. It is carcinogenic in hairless mice[17,18] and can produce distinctive changes in dermal connective tissue and in vasculature.[7,12,19-21] Importantly, when combined with UVB as in solar simulating radiation (SSR), it adds a significant contribution to the photodamage.[12] It has been recognized by some photobiologists that it is possible to raise chronic UVB exposure levels of hairless mice gradually to 4-6 MEDs without causing undue erythema whereas with SSR, the limit is ~1.5 MEDs (unpublished observations). Clearly, UVA has effects that need to be addressed when assessing photoprotection by sunscreens.

Yet another point to consider is that whereas the prevention of erythema by sunscreens is a given, based on the definition of SPF, mere prevention of sunburn may not be adequate during chronic exposures that continue over decades. It is well established that even with appropriate application and use of sunscreen products, small amounts of UVB and UVA penetrate to the viable layers of the skin. An SPF 15 sunscreen will block about 93% of the incident radiation. Sunburn may be prevented even after 4 or 5 hours of exposure, but the small amounts of transmitted radiation may have serious consequences with regard to chronic exposure.

The Rationale for This Study

To address these various issues, we were interested in examining the effect of transmitted radiation during chronic high dose exposure to SSR. Such studies are not feasible in humans, therefore we used the hairless mouse. On the basis of previous work to define an action spectrum for SSR-induced elastosis in this animal,[22] we determined that a cumulative dose of ~1 J/cm^2 of the UVB portion of the spectral emission of the solar simulator provides a cumulative UVA dose of ~170 J/cm^2. This combined dose, achieved over a period of 9 weeks, produced a 50% increase in elastic fibers as quantified by image analysis. Histologically, this was perceived as mild elastic fiber hyperplasia. For the purpose of this work, we defined the dose as a minimal photoaging dose (MPD).

To accomplish our aim, we irradiated hairless mice until 10 and 16 MPDs were accumulated. These cumulative doses were reached by exposing the mice twice weekly to 7 MEDs each for 18 and 30 weeks, respectively. The 7 MED dose was chosen to equal the SPF value of the basic sunscreen which contained only a UVB absorbing molecule. In addition, we tested an SPF 16 sunscreen containing the UVB absorber and a UVA II (320-340 nm) absorber and an SPF 18 sunscreen containing these two absorbers plus a UVA I (340-400 nm) absorber. These experimental conditions enabled us to examine the long-term consequences of UVA on UVB-protected skin, the effects of the small amounts of radiation transmitted through all sunscreens and the increase in protection provided by the addition of UVA absorbers to a UVB-absorbing sunscreen.

Materials and Methods

The materials and methods have been reported in detail elsewhere[23] and will therefore be described only briefly here.

Animals and Irradiation

Three groups of 18 albino hairless mice were irradiated twice weekly, nine at a time in a cage designed to prevent the mice from shielding each other. Room lighting (12 h on/off cycle) was with General Electric F-40 "GO" fluorescent tubes which emit no measurable UV radiation.

The radiation source was a forced air cooled 4200 W compact xenon arc solar simulator fitted with a 1 mm WG 320 Schott filter. The collimated beam passed through a water filter before impinging on a 45° UV-reflecting dichroic mirror. Irradiance at dorsal skin level was ~0.042 mW/cm^2 UVB and ~5.84 mW/cm^2 UVA. Spectral distribution was measured with an Optronic 742 spectral radiometer at the beginning and end of the study (R.M. Sayre, Rapid Precision Testing Labs, Covington, TN). Exposures of all groups over the first two weeks, were raised by 1/2 MED increments from 5 to 7 MEDs to avoid erythema in the group treated with the lowest SPF sunscreen. One MED was ~0.04 J/cm^2 as measured with a UVB sensor and required 18 minutes. This corresponds closely to an MED for a type II human under the same experimental conditions. To accumulate 10 and 16 MPDs at this rate required 18 and 30 weeks, respectively, with approximately 2 hour exposures twice weekly. Ambient temperature at mouse level was maintained at <27°C by the use of two small fans.

Sunscreens

Sunscreens were applied, just prior to irradiation, as 100 µl to cover the entire exposed dorsal surface of the mouse, providing ~2.5 $\mu l/cm^2$. The active ingredients of the three sunscreens were as follows: (1) octyl methoxycinnamate, mainly a UVB absorber (290-320 nm) with an SPF of 7; (2) the UVB absorber and oxybenzone with an SPF of 16, providing absorption of UVB and UVA II (290-340 nm); and (3) the UVB and UVA II absorbers plus avobenzone (Parsol 1789) a UVA I absorber (340-400 nm) with an SPF of 18 and providing full spectrum absorption (290-400 nm).

Neither unprotected nor vehicle controls were included because the 7 MED dose would not be tolerated under such circumstances. However, our large histologic data base includes skin from unirradiated, long-term UV irradiated at tolerated high doses and various vehicle controls which provide relevant comparisons.

Histology

Eight animals in each group were sacrificed at 10 MPDs (18 weeks) with the surviving remainder (8-9) at 16 MPDs (30 weeks). Dorsal skin biopsies (2 x 0.5 cm) were processed for light microscopy. The histochemical stains were: hematoxylin and eosin for general morphology, Masson's trichrome for collagen, Luna's aldehyde fuchsin for elastic fibers[24] and Mowry's colloidal iron for glycosaminoglycans.

Results

Histologic Findings

Many morphologic features of photodamaged skin were examined using the four histochemical stains. Because a complete descriptive account, along with multiple figures has appeared elsewhere[23] only the major findings will be reported here.

The most important overall observation is that, in the absence of erythema in all groups, some photodamage from transmitted radiation was seen regardless of the sunscreen used. As expected, the most damage occurred with the SPF 7 sunscreen, and the least with the SPF 18 sunscreen. Additionally, damage was greater at 16 MPDs than after 10 MPDs.

General Morphology (H & E Stain)

10 MPD

All specimens had moderate epidermal hyperplasia (6-8 cell layers vs 3-4 in normal controls) but in the SPF 18 group the epidermis was marginally thinner (5-7 cell layers). There was little or no cellular atypia in any of the groups. A few scattered sunburn cells were seen in the SPF 7 and 16 groups, but they were rare in the SPF 18 group.

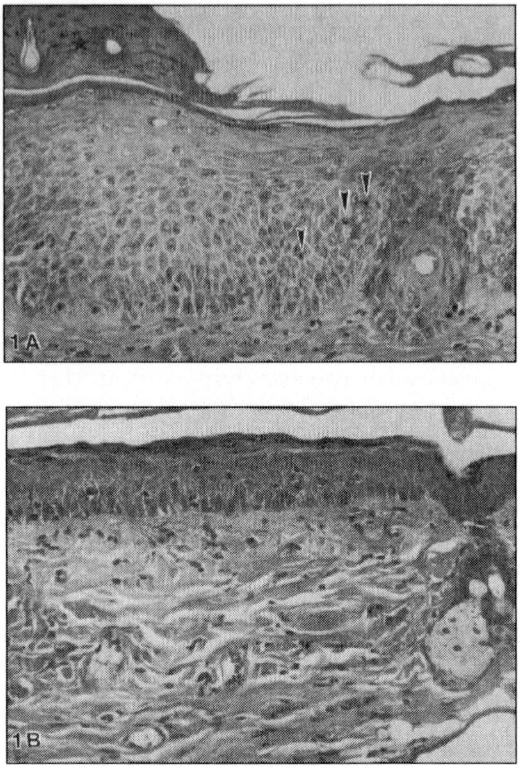

Fig. 5.1. 16 MPD: General Morphology. H & E stain. Magnification x295. (A) SPF 7 Sunscreen. The hyperplastic epidermis is characterized by thick downgrowths, loss of cellular order, atypia and numerous sunburn cells (▶). Foci of parakeratosis are present (*). (B) SPF 18 sunscreen. The moderately hyperplastic epidermis shows a mild lymphocytic infiltrate but little loss of order and no atypia.

16 MPD

Epidermal hyperplasia increased to 7-9 cell in the SPF 7 and 16 groups with many thicker downgrowths in the former (Fig. 5.1A) and fewer in the latter. Severe to moderate cellular atypia was present in the downgrowths. In the SPF 18 group there were 6-8 cell layers, very few downgrowths, and little or no atypia (Fig. 5.1B). Sunburn cells were considerably more abundant in the SPF 7 and 16 groups with only a few in the SPF 18 group.

Elastic Fibers (Luna's Stain):

10 MPD

With the SPF 7 and 16 sunscreens, elastic fiber hyperplasia was moderate and mild, respectively, whereas there was little to very mild hyperplasia with the SPF 18 sunscreen.

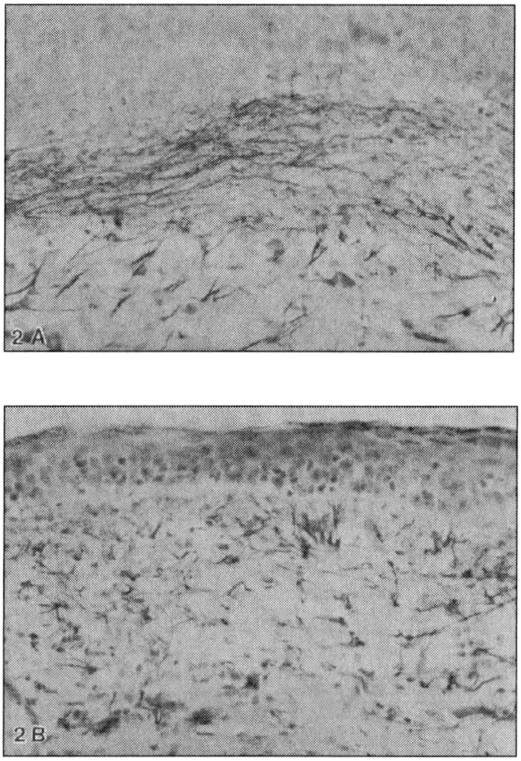

Fig. 5.2. 16 MPD: elastic fibers. Luna's Stain. Magnification x370. (A) SPF 7 sunscreen. Moderate focal elastic fiber hyperplasia. (B) SPF 18 sunscreen. Mild elastic fiber hyperplasia. Black and white reproduction has diminished the visible microscopic effects.

16 MPD

There was little change in the degree of hyperplasia except for a thickening of the fibers. The SPF 18 specimens showed the mildest changes (Fig. 5.2A,B).

Collagen (Masson's Tri-Chrome)

10 MPD

The only loss of affinity for the stain, an indication of collagen damage, was in the SPF 7 group.

16 MPD

There was increased loss of stain with the SPF 7 (Fig. 5.3A) and the appearance of some loss below epidermal downgrowths with SPF 16 whereas the SPF 18 group continued to show normal collagen staining (Fig. 5.3B).

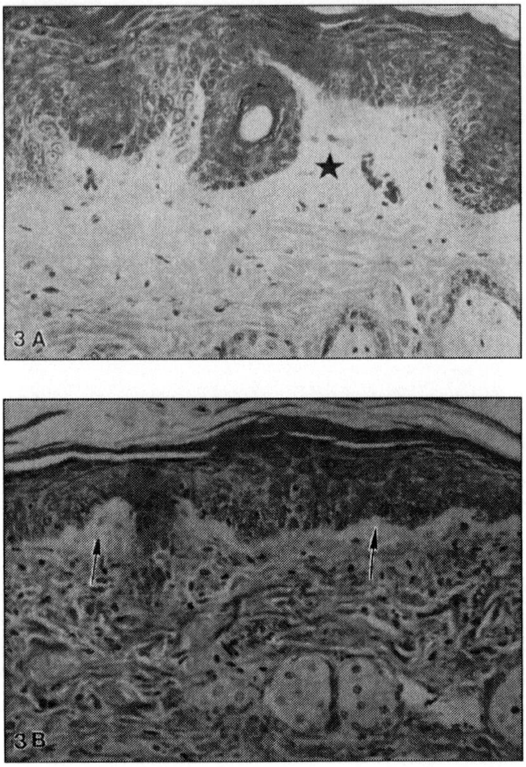

Fig. 5.3. (See color insert.) 16 MPD: Collagen. Masson's Tri-Chrome Stain. Magnification x295, (A) SPF 7 sunscreen. Loss of affinity for the stain is marked but is focal (*). (B) SPF 18 sunscreen. Collagen staining is normal. The pale subepidermal deposits (➔) are shown to be glycosaminoglycans when Mowry's stain is used.

Glycosaminoglycans (Mowry's Stain)

The most dramatic changes were in the development of the UV-induced glycosaminoglycans. For this reason, photomicrographs are shown for all groups at both dose points.

10 MPDs

Increases in the stippled blue staining glycosaminoglycans in the upper dermis were moderate and mild with SPF 7 and 16 sunscreen, respectively (Fig. 5.4 A,B). With the SPF 18 sunscreen a few specimens had one or two foci of very mild glycosaminoglycan deposits, but most had no increase (Fig. 5.4C).

16 MPDs

Glycosaminoglycans increased to severe with SPF 7, moderate with SPF 16, but there was still little or no increase with SPF 18 (Fig. 5.5A-C). In some of the specimens several foci of dense linear deposits of glycosaminoglycans were seen at the dermal-epidermal junction—a feature often associated with UVA radiation.

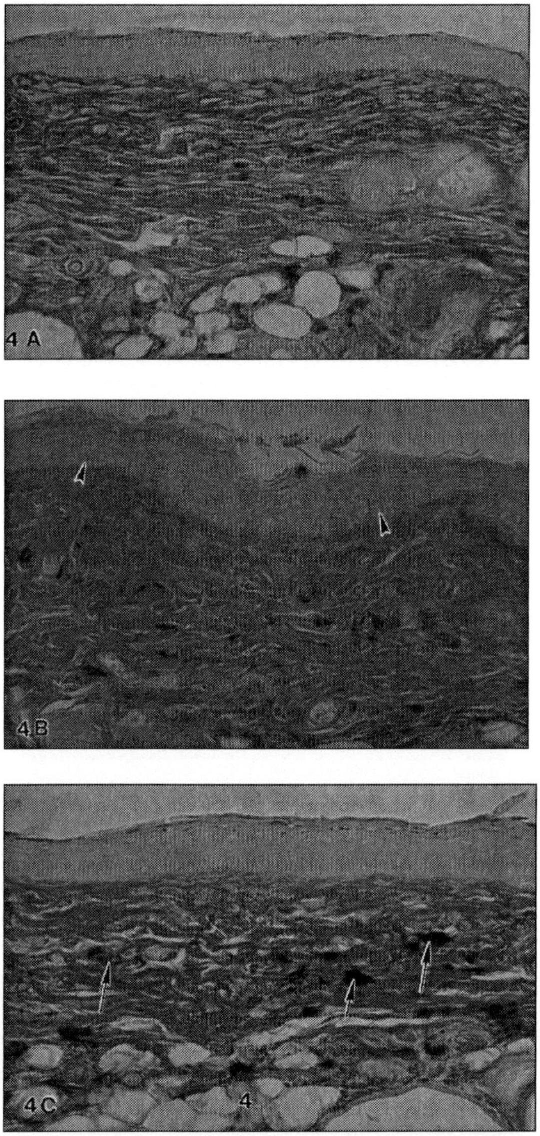

Fig. 5.4. (See color insert.) 10 MPD: Glycosaminoglycans (GAGs). Mowry's stain. Magnification x230. (A) SPF 7 Sunscreen. Extensive increases in blue staining GAGs are seen subepidermally. (B) SPF 16 Sunscreen. An area with barely discernible dermal GAGs but intercellular epidermal GAGs are present (▶). (C) SPF 18 Sunscreen. GAGs are not increased. The only blue staining dermal features are mast cells (➔).

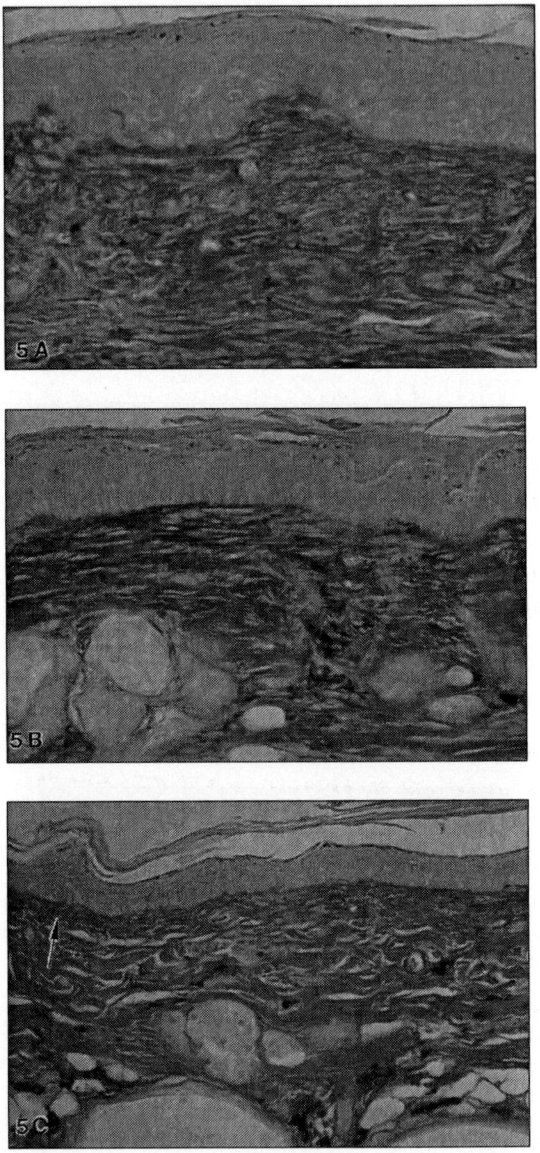

Fig. 5.5. (See color insert.) 16 MPD: Glycosaminoglycans (GAGs). Mowry's Stain. Magnification x230. (A) SPF 7 Sunscreen. Blue staining GAGs are considerably increased compared to 10 MPDs. (B). SPF 16 Sunscreen. GAGs are as abundant as with SPF 7 sunscreen. (C) SPF 18 sunscreen. GAGs are mildly increased with foci of dense deposits at the dermal-epidermal junction (→). All figures are reprinted from reference 23 with permission from coauthors Agin and Sayre.

Discussion

As this study demonstrates, with respect to the deleterious effects of chronic exposure to solar simulating radiation, it is not sufficient to merely prevent sunburn. Moreover, full spectrum protection is important. The UVB-absorbing sunscreen (SPF 7) allowed the greatest damage to the epidermis and connective tissue matrix, yet none of the animals developed erythema. Because the exposure dose (7 MEDs) approximately equaled the SPF, it is likely that the animal received an accumulation of transmitted radiation that was close to, but did not exceed 1 MED of SSR at each exposure. Much of the radiation reaching the skin with this sunscreen was UVA whereas about 14% of the UVB was transmitted. The unique UVB effects were evident in the focal loss of collagen staining, a histochemical feature that does not occur with UVA radiation even at very high doses.[12] The stippled pattern of glycosaminoglycan deposition was also characteristic of a UVB effect. Evidence for a UVA effect was seen by the occurrence of dense linear deposits at the dermal-epidermal junction.[12]

It should be made clear that the level of damage incurred in all sunscreen-protected animals was considerably less than that seen in unprotected control mice from our data base. These were exposed to ~1.5 MED of SSR thrice weekly for 30 weeks.[12] Elastosis was more severe, loss of collagen staining was extensive and glycosaminoglycans were greatly increased with a deposition pattern characteristic of both UVB and UVA exposure. Furthermore, in a previous photocarcinogenesis study, unprotected mice were exposed to 1 MED of SSR for only 10 weeks. Tumor incidence was 100% at week 27 with ~7 tumors/mouse.[25] By contrast, when the mice remaining at the 16 MPD point were examined, no tumors were present in the SPF 18 sunscreen. There was one small tumor low on the flank of one animal in the SPF 16 group and with the SPF 7 sunscreen only 3 of the 9 surviving mice had one or two small tumors at the end of the 30 week study.[23] Thus, even the low SPF sunscreen provided significant protection against tumorigenesis.

The present study suggests that dermal connective tissue is especially vulnerable to the small amounts of transmitted radiation. Evolution of the damage appears to be more rapid than that of carcinogenesis. Protection was clearly increased with the SPF 16 sunscreen which contained, in addition to the UVB absorber, a UVA II absorber (320-340 nm) and an erythema protection factor more than twice that of the UV dose delivered. The SPF 18 sunscreen, with the addition of the UVA I absorber (340-400 nm: peak absorption 355 nm) provided the most effective protection. As would be expected, photodamage in all groups was greater at 16 MPDs than at 10 MPDs. However, there was less progression of the damage with time and greater MPDs in the SPF 18 sunscreen group than in the other two groups.

Deleterious effects of suberythemal doses of UV have been reported in human studies. Our observation of sunburn cells (SBCs) in most specimens confirms findings in acute studies by Kaidbey.[2] After treatment with SPF 15 and 30 sunscreens, subjects were exposed to ~15 MEDs of SSR. Faint erythema developed in some of the SPF-15 protected sites but not in the SPF 30 sites. Although the greatest number of SBCs occurred with the former, SBCs were induced in all subjects and with both sunscreens. Thus, injury can also occur in human epidermal cells in the absence of erythema and with doses that are far below the SPF of the sunscreen. Similarly, Pearse et al,[1] treating human subjects with an SPF 5 sunscreen and irradiating with 2, 4 and 6 MEDs of UVB, found erythema to be prevented in the 2 and 4 MED groups, but skin thickening and epidermal enzyme activity were affected by the transmitted radiation.

Using a different approach, Lavker et al[26] exposed human subjects, at different sites, to 0.5 MED of SSR (290-400 nm) and 0.5 MED of UVA (320-400 nm) over a period of 5 weeks for 28 doses. Compared to unirradiated skin, major changes were: increased binding of lysozyme to elastic fibers, suggesting some change in the fibers and increased inflammation as determined by leukocyte common antigen deposition. Similar changes were produced by both spectra with the greatest effect seen with UVA.

Our findings and those of Lavker et al[26] add to the growing evidence that although UVB is the most damaging portion of terrestrial sunlight, UVA is responsible for numerous morphological, molecular and biochemical events which contribute to photodamage of skin.[12,20,27,28] Moreover, the effects of long-term UVA radiation are qualitatively and quantitatively different from those of UVB.[12,19,21]

Despite the mounting evidence from these and other studies, it has been argued that the amount of UVA radiation transmitted through high SPF sunscreens is not a significant factor in chronic photodamage.[29] This theoretical conclusion was reached with calculations based on acute exposure and erythema as an endpoint.

This fails to consider the many biological consequences to all skin structures of chronic exposure over decades. Also not considered is the relatively low absorbency characteristics of some UVA II absorbers, the absence of UVA I absorbers in most sunscreens and the long-term effects of full spectrum UVA on dermal matrix damage. For a UVA protection factor to be such that it is capable, mainly, of preventing UVA, induced erythema is not supported by in vivo experiments which show skin damage at suberythemal UVA doses. The claim that skin damage from 150 minutes of protected exposure with an SPF 15 sunscreen would be the same as from 10 minutes of unprotected exposure[29] is therefore true only for erythema.

On the other side of the argument, pointing out the need for broad spectrum sunscreens, Diffey made a revealing calculation on the UVA dose received with and without sunscreens on a clear day at the beach.[30] Using 20 minutes as the approximate time to receive 1 MED without a sunscreen, Diffey reports that 6 J/cm^2 of UVA would be absorbed by the skin. With an SPF 8 sunscreen containing a UVB absorber and a UVA absorber such as the benzophenones, a subject could remain in the sun for 2.5-3 hours without developing erythema. However, the skin would receive a total of 15 J/cm^2 of UVA. Interestingly, this is quite close to the suberythemal dose used by Lavker et al.[26]

Thus, it is becoming increasingly clear that with "real life" conditions such as full spectrum solar radiation, chronic exposure for several hours in the sun and the inevitable radiation transmitted through sunscreens, the mere prevention of erythema does not totally prevent photoaging. As this study demonstrates, the sunscreen with the highest SPF and the broadest absorption characteristics gave the best protection against suberythemal transmitted SSR radiation. In addition, compared to the lower SPF sunscreens, the SPF 18 sunscreen reduced the progression of damage over time as the SSR dose increased from 10 MPD to 16 MPD. These data strongly support the need for broad spectrum sunscreens that include both UVA II and UVA I absorbing molecules as well as SPFs greater than 15 if the accumulation of photodamage over the many decades of human sun exposure is to be moderated.

References

1. Pearse AD, Marks R. Response of human skin to ultraviolet radiation: Dissociation of erythema and metabolic changes following sunscreen protection. J Invest Dermatol 1983; 80:191-194.

2. Kaidbey KH. The photoprotective potential of the new superpotent sunscreens. J Am Acad Dermatol 1990; 22:449-452.
3. Kligman LH. The hairless mouse and photoaging. Yearly Review. Photochem Photobiol 1991; 54:1109-1118.
4. Snyder DS, May M. Ability of PABA to protect mammalian skin from ultraviolet light-induced skin tumors and actinic damage. J Invest Dermatol 1975; 65:543-546.
5. Kligman LH, Akin FJ, Kligman AM. Sunscreens prevent ultraviolet photocarcinogenesis. J Am Acad Dermatol 1980; 3:30-35.
6. Kligman LH, Akin FJ, Kligman AM. Prevention of ultraviolet damage to the dermis of hairless mice by sunscreens. J Invest Dermatol 1982; 78: 181-189.
7. Bissett DL, Hannon DP, Orr TV. An animal model of solar aged skin: Histologic, physical and visible changes in UV-irradiated hairless mouse skin. Photochem Photobiol 1987; 46:367-378.
8. Kligman AM, Kligman LM. Photoaging. In: Fitzpatrick TB, Eisen AZ, Wolff K et al, eds. Dermatology in General Medicine. 4th ed. McGraw Hill, Inc. 1993:2972-2979.
9. Plastow SR, Harrison JH, Young AR. Early changes in dermal collagen of mice exposed to chronic UVB irradiation and the effects of a UVB sunscreen. J Invest Dermatol 1988; 91:590- 592.
10. Fourtanier A, Labat-Robert J, Kern P et al. In vivo evaluation of photoprotection against chronic ultraviolet-A irradiation by a new sunscreen Mexoryl SX. Photochem Photobiol 1992; 55:549-560.
11. Kligman LH, Zheng P. The protective effect of a broad spectrum sunscreen against chronic UVA radiation in the hairless mouse. A histologic and ultrastructural assessment. J Soc Cosmet Chem 1994; 45:21-33.
12. Kligman LH, Akin FJ, Kligman AM. The contributions of UVA and UVB to connective tissue damage in hairless mice. J Invest Dermatol 1985; 84:272-276.
13. Van Praag MCG, Roza L, Boom BW et al. Determination of the photoprotective efficacy of a topical sunscreen against UVB-induced DNA damage in human epidermis. J Photochem Photobiol B: Biol 1993; 19:129-134.
14. Whitmore SE, Morison WL. Prevention of UVB-induced immunosuppression in humans by a high sun protection factor sunscreen. Arch Dermatol 1995; 131:1128-1133.
15. Roberts LK, Beasley DC. Commercial sunscreen lotions prevent ultraviolet radiation-induced immune suppression of contact hypersensitivity. J Invest Dermatol 1995; 105:339-344.
16. Kligman LH, Akin FJ, Kligman AM. Sunscreens promote repair of ultraviolet radiation-induced dermal damage. J Invest Dermatol 1983; 81:98-101.
17. Kelfkens G, de Gruijl FR, van der Leun JC. Tumorigenesis by shortwave ultraviolet A: papillomas versus squamous carcinomas. Carcinogenesis 1991; 12:1377-1382.

18. Kligman LH, Crosby MJ, Miller SA et al. Carcinogenesis by long wavelength UVA and failure of reciprocity. In: Urbach F, ed. Biological Responses to Ultraviolet A Radiation. Valdenmar Publishing Co., 1992:301-308.
19. Zheng P, Kligman LH. UVA-induced ultrastructural changes in hairless mouse skin: A comparison to UVB-induced changes. J Invest Dermatol 1993; 100:194-199.
20. Kligman LH, Gebre M. Biochemical changes in hairless mouse skin collagen after chronic exposure to UVA radiation. Photochem Photobiol 1991; 54:233-237.
21. Kligman LH. UVA-induced biochemical changes in hairless mouse skin collagen: a contrast to UVB effects. In: Urbach F, ed. Biological Responses to Ultraviolet A Radiation. Valdenmar Publishing Co., 1992:209-215.
22. Kligman LH, Sayre RM. An action spectrum for ultra-violet induced elastosis in hairless mice: Quantification of elastosis by image analysis. Photochem Photobiol 1991; 53:237-242.
23. Kligman LH, Agin PP, Sayre RM. Broad spectrum sunscreens with UVA I and UVA II absorbers provide increased protection against solar simulating radiation-induced dermal damage in hairless mice. J Soc Cosmet Chem 1996; 47:129-155.
24. Kligman LH. Luna's technique: a beautiful stain for elastin. Am J Dermatopath 1981; 3:199-201.
25. Kligman LH, Crosby MJ. Topical tretinoin enhances corticosteroid-induced inhibition of tumorigenesis in hairless mice previously exposed to solar simulating radiation. Cancer Letters 1996; 107:217-222.
26. Lavker FM, Gerberick GF, Veres D et al. Cumulative effects from repeated exposures to suberythemal doses of UVB and UVA in human skin. J Amer Acad Dermatol 1995; 32:53-62.
27. Scharffetter K, Wlaschek M, Hogg A et al. UVA radiation induces collagenase in human dermal fibroblasts in vitro and in vivo. Arch Dermatol Res 1991; 283:506-511.
28. Wlaschek M, Bolsen K, Herrmann G et al. UVA-induced autocrine stimulation of fibroblast-derived-collagenase by IL-6; a possible mechanism in dermal photodamage? J Invest Dermatol 1993; 101:164-168.
29. Urbach F. Ultraviolet A transmission by modern sunscreens: is there a real risk? Photodermatol Photoimmunol Photomed 1992/1993; 9:237-241.
30. Diffey BL. The need for sunscreens with broad spectrum protection. In: Urbach F, ed. Biological Responses to Ultraviolet A Radiation. Valdenmar Publishing Co., 1992:321-328.

CHAPTER 6

A Transgenic Mouse Model of Cutaneous Photoaging Measuring Elastin Promoter Activation

Eric F. Bernstein

Intrinsic Versus Extrinsic Aging

Cutaneous aging consists of two distinct processes: chronological or intrinsic again, which refers to the biological clock which keeps ticking regardless of outside influences, and sun-induced aging which reflects our exposure to ultraviolet radiation as well as our skin's ability to protect itself from damaging ultraviolet rays. Chronological aging of skin is a gradual process and results in a slow decrease in dermal thickness in elderly individuals.[1-3] In addition, the cellularity of the dermis decreases, resulting in fewer fibroblasts, mast cells, and blood vessels in chronologically aged skin.[1] In addition, there is an age-associated loss of normal elastic fibers, which may contribute to some of the fine wrinkling seen in sun-protected, chronologically aged skin. The elastic fiber network is responsible for maintaining the skin's ability to return to its original shape and position after physical distention. With progressing chronological age, the mechanical properties of skin change, reflecting a loss of elastic recovery.[4,5] This reflects the altered functional ability of the elastic fiber network in aged skin. It has also been shown that collagen and collagenase gene expression decreases in fibroblasts taken from more aged donors. In addition, the genes for type I collagen and collagenase were less responsive to modulation by pro-inflammatory cytokines in fibroblasts from more aged individuals.[6] Thus, alterations in collagen and collagenase accompany elastic fiber alterations, and contribute to the changes observed in sun-protected, chronologically aged skin.

Sunscreen Photobiology: Molecular, Cellular and Physiological Aspects, edited by Francis P. Gasparro.

Extracellular Matrix

Thus, most of the visible alterations occurring in photoaged skin are the result of changes in the dermal extracellular matrix. The extracellular matrix is comprised of all the intercellular material which makes up the bulk of the dermis, with the exception of water. This includes mostly proteins and complex sugars arranged in a very organized structure to serve the numerous functions necessary to support living skin. The main components of the dermis include proteins such as collagen and the proteins comprising elastic fibers, proteoglycans, and glycosaminoglycans. Most of the extracellular matrix comprising the dermis is synthesized by dermal fibroblasts, which are influenced by numerous factors, including cytokines, inflammatory cells, and extracellular matrix components themselves. Physical factors such as sunlight also play an important role in directing the deposition of extracellular matrix. In human skin, the main extracellular matrix components are the proteins comprising collagen and elastic fibers, and the complex sugars called glycosaminoglycans or proteoglycans when complexed with a protein core. The fibrillar collagens form the majority of the dermal matrix and provide skin with both substance and strength to resist outside forces. Collagen is the most abundant of the extracellular matrix components and its name is derived from the Greek (*kolla*, glue; *genna*, to produce; collagen, glue former), because when collagen is boiled, it produces glue. The fibrillar collagens types I and III are the most abundant in skin. Type I collagen makes up approximately 80% of the collagen found in skin and is one of the major structural components of the human body.[7] It is comprised of three polypeptide chains which are arranged in a triple helical configuration. Two of the chains making up type I collagen are identical and are labeled $\alpha 1$(I) chains, while the third chain differs from the first two and is named $\alpha 2$(I).[8-10] Type III collagen, also referred to as fetal collagen because of its abundance in fetal tissues, is the second major fibrillar collagen comprising the human dermis.[11] Although it accounts for approximately half of the total collagen in fetal skin, production of type I collagen then exceeds that of type III collagen and the resulting ratio of type I to type III collagen is about 6:1 in adult skin.[12] Type III collagen is also quite abundant in other extensible tissues including the gastrointestinal tract and the arterial blood vessels. Type III collagen is composed of three identical chains ($\alpha 1$(III)) and is arranged in the typical triple helical conformation characteristic of fibrillar collagens.

Although elastic fibers comprise only about 1-2% of the weight of normal skin after removal of water, they are crucial to providing elasticity (the ability to stretch) and resilience (the ability of skin to return to its original shape after stretching) to skin.[13-15] Elastic fibers are composed of two proteins, elastin and fibrillin. Fibrillin represents the more stringy or fibrillar component of the elastic fiber. One might imagine a bungee cord with the stretchy inner component which represents elastin, and the surrounding cloth string which would represent the fibrillin. Elastic fibers form a plexus in the papillary dermis, extending vertically in the dermal papillae and surrounding superficial dermal blood vessels. In the reticular dermis, elastic fibers are much thicker and run parallel to the epidermis in between and surrounding the larger collagen fibers. Elastic fibers also surround the adnexal structures of the deeper dermis. A possible explanation for the ability of elastic fibers to recoil after stretching is that the hydrophobic domains are exposed to the aqueous extracellular milieu upon stretching. Contraction then ensues by folding of elastin into its original conformation, in which the hydrophobic residues reside in the interior, away from the aqueous environment.[16]

Glycosaminoglycans are polysaccharides which are composed of repeating disaccharide units. The disaccharides are ringed sugars, and each different glycosaminoglycan has a slightly different structure to these repeating units. They are widely distributed throughout skin but comprise only a small amount of the weight of skin after removal of water. However, these glycosaminoglycans bind a substantial amount of water relative to their weight, and therefore provide hydration to the skin. Hyaluronic acid is one of the most abundant glycosaminoglycans in skin, both in the dermis and the epidermis, and may be present either bound to a protein core, or unbound. The other glycosaminoglycans found in skin are linked to a protein by the terminal reducing sugar residue, thus forming proteoglycans.[17,18] Proteoglycans consist of one or a few glycosaminoglycans linked to a protein core. Glycosaminoglycans and proteoglycans serve a variety of functions, which include promoting cell movement within tissues, maintenance of basement membrane integrity, regulation of cellular interactions, and regulation of collagen and possibly elastic fiber formation. Two proteoglycans, decorin and versican, may be involved in collagen and elastic fiber formation, respectively.[19-21]

Photoaging

The effects of chronic sun exposure are easily identifiable when comparing sun-exposed skin to sun-protected sites on the same individual. Superimposed on the sagging and fine wrinkling of chronologically aged skin are the deep furrows, pigmentary alterations, benign and malignant growths, and thickened leathery appearance of chronically sun-damaged skin. Most of the characteristics we typically associate with an aged appearance result not from intrinsic aging, but from chronic photodamage.[22-23] Histopathologic examination of sun-exposed skin reveals the accumulation of large amounts of material that stains similarly to elastin when examined on routine hematoxylin-eosin stained sections, and is thus termed solar elastosis.[22,23,24-28] This material also stains strongly with elastic tissue stains such as Verhoeff-van Gieson stain. Immunohistochemical analysis of sun-exposed skin has shown that solar elastosis is composed of elastin[29-31] fibrillin,[29,31,32] versican,[33] and hyaluronic acid,[34] the normal constituents of elastic fibers. Although these fibers are composed of the same constituents as normal elastic fibers, they are present in greatly increased amounts and arranged in an abnormal and haphazard manner (Fig. 6.1). Biochemical analysis of sun-damaged skin has revealed an elevated concentration of desmosines (crosslinked compounds found only in elastic fibers) in sun-damaged skin.[35]

Accompanying the alterations in elastic fiber formation are changes in the amount and organization of collagen fibers within the area of solar elastosis in sun-damaged skin. Studies have suggested that degradation of collagen takes place in areas of solar elastosis.[36] Direct assays of collagen content[37] and immunohistochemical staining for collagen[29,37-39] have revealed changes in the collagen of photodamaged sites. Most studies have shown that the amount of collagen is reduced in areas of solar elastosis, a finding which may be indicative of collagen degradation and/or alterations in collagen production. Thus, alterations in steady-state collagen levels accompany the deposition of abnormal elastic fibers which characterizes sun-damaged skin.

Alterations of glycosaminoglycans and proteoglycans are also part of the changes taking place as a result of chronic sun-exposure. Glycosaminoglycans have been shown to be increased in the region of solar elastosis in chronically sun-exposed skin.[40] Recently, the large chondroitin sulfate proteoglycan, versican, has been shown to

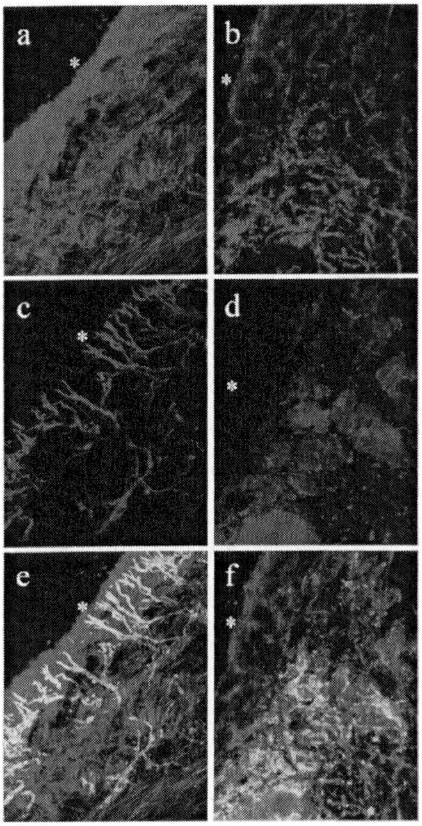

Fig. 6.1. (See color insert.) Laser scanning confocal microscopy of normal and sun-damaged skin from the same individual. (a) Elastic fibers normally form a delicate meshwork of fibers which extends perpendicularly from the dermal-epidermal junction in sun-protected skin. (b) Chronically photodamaged skin contains numerous, very large, clumped elastic fibers in the superficial dermis, replacing the normal architecture. (c) Sun-protected skin also reveals strongly staining collagen just beneath the dermal-epidermal junction. Below this is a dense network of fibers. (d) Sun-damaged skin also contains a band of strongly staining collagen. However, the collagen below this area is stringy and much less dense than sun-protected skin. (e) Double-labeled images show the interrelationship between collagen and elastic fibers. Collagen is red and elastin appears green or yellow. Sun-protected skin demonstrates an orderly arrangement of fibers. (f) In contrast with protected skin, sun-damaged skin reveals a haphazard arrangement of the fibers. Asterisk indicates dermal-epidermal junction in e and f *Bar*, 10 μm. Reprinted with permission from reference 39.

codistribute with elastic fibers in human skin.[41] In addition, the small chondroitin sulfate proteoglycan, decorin, codistributes with collagen fibers in the human dermis.[42-46] Versican and hyaluronic acid have been found to be present on both normal elastic fibers and in the abnormal solar elastotic material characteristic of sun-damaged skin.[33,34] Thus at least at the light microscopic level, the solar elastotic material seems to be composed of the constituents comprising the normal elastic fibers found in sun-protected skin.

Animal Model of Cutaneous Photoaging Measuring Elastin Promoter Induction

Although several components of the extracellular matrix have been found to be altered in photodamaged skin, deposition of solar elastotic material is by far the most prominent change present in the dermis of sun-damaged skin. Previous studies have shown increases in elastin mRNAs accompanying the increase in elastin in sun-damaged sites. This finding has been demonstrated in fibroblasts derived from skin biopsies taken from sun-damaged and sun-protected sites in the same individual, and measured by direct extraction of RNA from biopsy samples.[31] In addition, transient transfections of fibroblast cultures derived from sun-protected versus sun-damaged sites with an elastin promoter-CAT construct, revealed an approximately 4-fold increase in elastin promoter activity in cells derived from sun-damaged sites.[31] Quantitatively, this 4-fold increase correlates well with the previously described 4-fold increase in desmosine content of photodamaged skin.[15]

To develop a rapid and sensitive means of studying photoaging and compounds which may protect against sun-damage in vivo, we utilized a transgenic mouse line containing the human elastin promoter linked to a chloramphenicol acetyltransferase (CAT) reporter gene.[47,48] Because promoter induction is one of the primary events leading to elastin production in skin,[31] this model permits precise and rapid measurements of one of the earliest events leading to cutaneous photoaging. These mice contain a portion of the human elastin gene which is responsible for regulating elastin production, by either increasing or decreasing elastin mRNA production, in response to ultraviolet radiation. They are phenotypically normal and indistinguishable from mice of the same strain not containing the transgene. However, the elastin promoter is connected to a CAT gene, which reports on the level of elastin promoter activation by production of an enzyme not normally found in mammalian cells (CAT). By obtaining a skin biopsy sample, or harvesting fibroblasts derived from the skin of these mice, the level of elastin gene induction may be indirectly determined when assayed for CAT activity. This system has several advantages over other methods of measuring photoaging in a murine model. For example, the results are quantitative, and large increases in elastin promoter activity are observed after even a single exposure to UV.[48] In addition, these mice contain the human elastin gene and thus may more closely simulate the human response to UV, however, this system still reflects differences between the human and murine inflammatory response to UV as well as the thinness of mouse skin as compared human skin. Most importantly, this assay requires days and as few as a single UV exposure, in contrast to the months of continuous exposures required to generate even modest amounts of solar elastosis.

We exposed the mice, as well as fibroblast cultures derived from skin explants from these mice, to ultraviolet B radiation (UVB) and ultraviolet A radiation (UVA). Our results show that a large increase in elastin promoter activity results from irradiation with UVB both in vivo and in vitro, while a more modest increase occurs in response to UVA in vivo without a corresponding in vitro response (Figs. 6.2a and b).[48] In addition, modest amounts of UVB were shown to result in large increases in elastin gene induction, while relatively large amounts of UVA demonstrated no increase. (Figs. 6.2c and d, Fig. 6.3).[48] Thus, both UVA and UVB activate the human elastin promoter in transgenic mice, while only UVB and not UVA has this effect in vitro. These data suggest that inflammatory cells may not be necessary for elastin promoter activation by UVB, and suggests that a direct effect of UVB on dermal fibroblasts is in large part responsible for elastin promoter activation. In addition, cells other than dermal fibroblasts such as resident mast cells, inflammatory cells, or keratinocytes elaborating cytokines, may be the primary target for elastin gene induction by UVA.

Having demonstrated the ability to measure elastin gene induction in a rapid and sensitive manner, this system was then used to measure the protection against elastin promoter activation afforded by sunscreens having various sun protection factor (SPF) ratings. Sunscreens previously shown to have SPF ratings of 2,4,8, and 15 were applied to the backs of mice prior to the administration of solar simulating radiation (SSR). Irradiated control mice received the lotion vehicle used for all sunscreen preparations. Control mice received no lotion or irradiation. Sunscreens or the lotion vehicle alone produced no alterations in CAT activity. Sunscreens were shown to protect against SSR, with a linear increase in protection with increasing SPF of the sunscreen (Fig. 6.4).[49]

Although the large increase in CAT activity in response to UVB (or SSR) permits a high degree of sensitivity in evaluating compounds thought to afford protection against UVB, the modest increase in CAT activity measured in response to UVA in vivo and the absence of a response to UVA in vitro make evaluation of agents which may protect against UVA more difficult. To determine if the addition of 8-methoxypsoralen (8-MOP) to our system would result in promoter activation by UVA in vitro and enhance the promoter response in vivo, and to develop a model to study compounds thought to protect against UVA, we examined the effects of PUVA in our transgenic

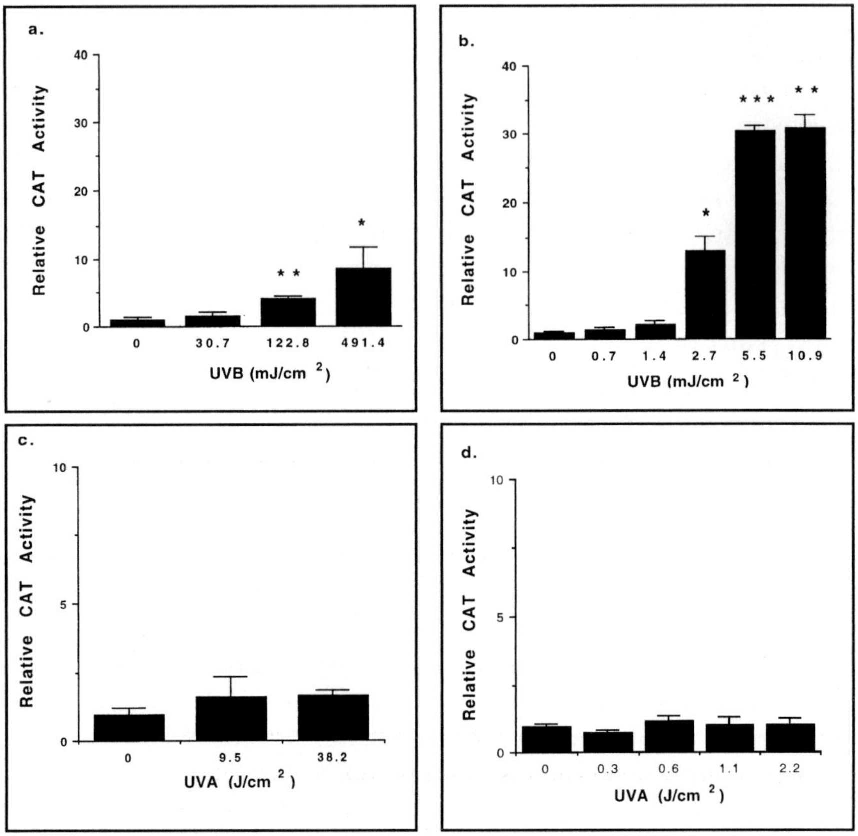

Fig. 6.2. UVB induces greater increases in elastin promoter activity than UVA. (a) CAT activity responds in a dose-dependent fashion to UVB in mice. Single doses of 30.7, 122.8, and 491.4 mJ/cm² of UVB administered over 3.75, 15 and 60 minutes, respectively, resulted in increasing CAT activities up to 8.5-fold that of unirradiated control mice (0). The lowest dose given corresponds to approximately 1.5 human MEDs of UVB. (b) Cutaneous fibroblasts derived from transgenic mice were exquisitely sensitive to UVB, with maximal promoter activation over 30-fold higher than unirradiated controls (0) after a single dose of only 5.5 mJ/cm². (c) CAT activity increases in response to UVA in vivo. Single doses of 9.5 and 38.2 J/cm² of UVA administered over 1.3 and 5.2 hours, respectively, increased CAT activity less than 2-fold over unirradiated controls (0). This difference approached statistical significance ($p=0.07$). The smallest dose given represents approximately one-fourth of a human MED for UVA. (d) UVA irradiation fails to increase CAT activity in vitro. Although substantial doses of UVA were administered over as long as 18.4 minutes (2.2 J/cm²), CAT activity of fibroblast cultures was virtually unchanged, as compared to unirradiated controls (0). The failure of UVA to increase promoter activity in vitro, despite achieving almost 2-fold increases in vivo, suggests that the primary target for UVA-induced elastin promoter activation is a cell other than the dermal fibroblast. Thus, the mechanism for elastin promoter activation by UVA differs qualitatively from that of UVB. *Bars* indicate mean + SD. Statistical significance is indicated above *bars* (* $p<0.05$; ** $p<0.01$; *** $p<0.001$). Reprinted with permission from reference 48.

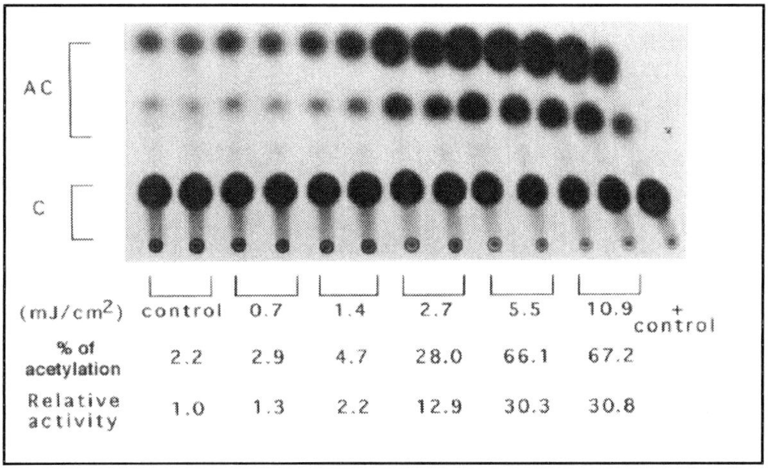

Fig. 6.3. A small single dose of UVB radiation causes large increases in CAT activity in vitro. Fibroblasts established from the skin of transgenic mice responded in a dose-dependent fashion to small doses of UVB. One-fourth of a human MED, administered over 40 seconds in a single dose, increased elastin promoter activity over 30-fold. These data suggest that elastin promoter activation by UVB results, at least in part, from direct effects on dermal fibroblasts and is independent of an inflammatory response. Reprinted with permission from reference 48.

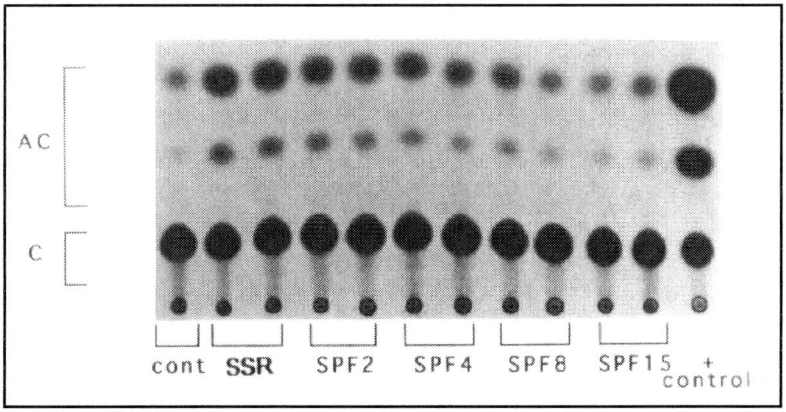

Fig. 6.4. Sunscreens protect against photoaging as measured by elastin promoter activation in transgenic mice. Mice were pretreated topically with either a vehicle alone (SSR), a sunscreen with an SPF of 2 (SPF 2), an SPF of 4 (SPF 4), an SPF of 8 (SPF 8), or an SPF of 15 (SPF 15), followed by 30 human MEDs of SSR. Untreated control mice (cont.) received no topical treatment or SSR. Elastin promoter activity increased approximately 6-fold in response to SSR. Increasing amounts of protection were afforded by sunscreens with increasing SPFs. The positive control represents an extract from the skin of a transgenic mouse previously treated with UVB, and known to result in elastin promoter upregulation. Each lane represents an assay from a separate mouse. Reprinted with permission from reference 49.

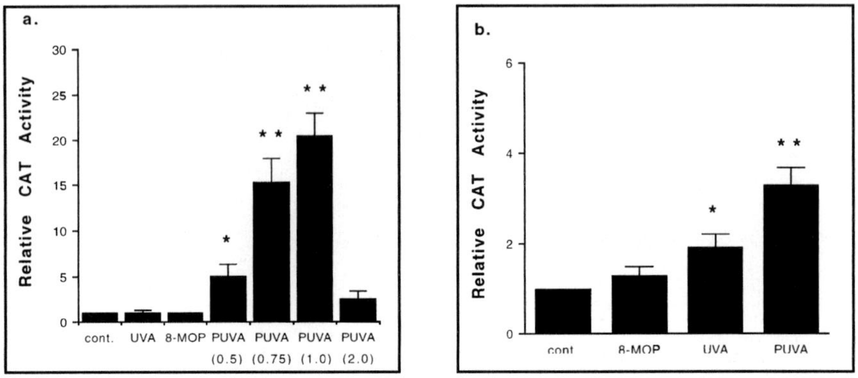

Fig. 6.5. The addition of 8-MOP increases promoter activation in response to UVA. (a) The average increase in CAT activity in vitro, including all experiments, reached almost 20-fold that of untreated controls. UVA doses of 0.5, 0.75, 1.0 and 2.0 J/cm² administered to cells preincubated with 1.0 mg/ml of 8-MOP resulted in elastin promoter activation of 2.2-fold 1.1 (mean ± sem), 15.5-fold ± 5.2, 19.8-fold ± 2.2, and 2.5-fold ± 1.9, respectively. *Bars* indicate mean ± sem. Statistical significance is indicated above *bars* (* $p<0.05$; ** $p< 0.01$). (b) The average increase in CAT activity in vivo, including all experiments, reached 3.3-fold that of untreated controls. 8-MOP alone (8-MOP) resulted in an insignificant increase in CAT activity of 1.3-fold as compared to untreated controls (cont.). UVA alone (UVA) increased CAT activity almost 2-fold, a statistically significant increase. The addition of 8-MOP to UVA-treated mice increased CAT activity over 70% more than UVA treatment alone (PUVA). The average increase of 3.3-fold in the PUVA treated mice was statistically significant. *Bars* indicate mean ± sem. Statistical significance is indicated above *bars* (* $p<0.05$; ** $p< 0.01$) Reprinted with permission from reference 50.

mouse model. The addition of 8-MOP to this system resulted in large increases in CAT activity both in vitro and in vivo, which were comparable to those resulting from UVB (Fig. 6.5).[50] This provided a model for studying the effect of agents which protect against UVA. To test the sensitivity of this system, sunscreens affording increasing amounts of UVA protection were evaluated in vivo. Sunscreens with increasing amounts of UVA protection afforded increasing protection against elastin promoter activation in vivo (Fig. 6.6).[51]

The sensitivity and rapid nature of this assay enables the quick screening of agents which may protect against UV-induced damage. Combining the use of fibroblast cultures derived from the skin of transgenic mice with studies using the skin of transgenic mice, may help elucidate the factors responsible for photoaging, and better determine the mechanisms by which various agents afford protection against UV. Future studies to determine the mechanisms by which extracellular matrix alterations occur in photoaging, should lead to new strategies for preventing the deleterious effects of chronic sun exposure.

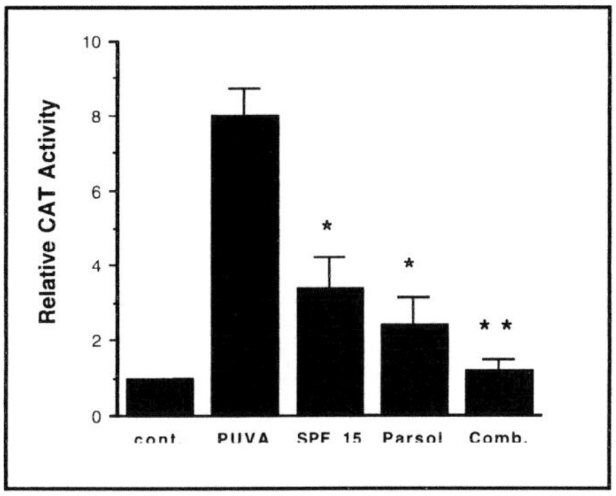

Fig. 6.6. Sunscreens protect against photoaging as measured by elastin promoter activation in transgenic mice. Mice were pretreated topically with 8-MOP prior to sunscreen administration. Mice received either a vehicle alone containing no sunscreen (PUVA), a sunscreen with an SPF of 15 having some absorption in the UVA (SPF 15), a sunscreen with significant absorption in the UVA range (Parsol), and a combination product combining the ingredients of the SPF 15 sunscreen and the UVA filter (Comb.). Untreated control mice (cont.) received no topical treatment or UVA. Elastin promoter activity increased 8-fold in response to PUVA. Increasing amounts of protection were afforded by the SPF 15 sunscreen, the UVA filter, and the combination product, respectively. The positive control represents an extract from the skin of a transgenic mouse previously treated with UVB, and known to result in elastin promoter upregulation. Each lane represents an assay from a separate mouse. Reprinted with permission from reference 51.

References

1. Gilchrest BA. Aging of Skin. In: Fitzpatrick TB, Eisen AZ, Wolff K, Freedberg IM, Austen KF, eds. Dermatology in General Medicine. Vol. I. New York: McGraw Hill 1993.
2. Tan CY et al. Skin thickness measurement by pulsed ultra-sound: Its reproducibility, validation and variability. Br J Dermatol 1982; 106:657.
3. de Rigal J et al. Assessment of aging of the human skin by in vivo ultrasonic imaging. J Invest Dermatol 1989; 93:621.
4. Daly CH, Odland GF. Age-related changes in the mechanical properties of human skin. J Invest Dermatol 1979; 73:84.
5. Escoffier C et al. Age-related mechanical properties of human skin: An in vivo study. J Invest Dermatol 1989; 93:353.
6. Chen YQ, Mauviel A, Uitto J. Age-related changes in the expression and cytokine response of extracellular matrix genes in human dermal fibroblast cultures. J Geriatr Dermatol 1994; 2:163-169.
7. Uitto J. Molecular pathology of collagen in cutaneous diseases. In: Advances in Dermatology. St. Louis: Mosby Year-Book, Inc., 1990:313-328.
8. Kuivaniemi H, Tromp G, Chu M-L, Prockop DJ. Structure of full-length cDNA clone for the pre pro-a2(I) chain of human type I procollagen. Biochem J 1988; 252:633-640.
9. Chu M-L, Mann K, Deutzmann R, Pribula-Conway D, Hsu-Chen CC, Bernard MP, Timpl R. Characterization of three constituent chains of collagen type VI by peptide sequences and cDNA clones. Eur J Biochem 1987; 168:309-317.

10. Myers JC, Loidl HR, Seyer JM, Dion AS. Complete primary structure of the human α2 type V procollagen COOH-terminal pro-peptide. J Biol Chem 1985; 260:2315-2320.
11. Epstein EH. $\alpha 1(III)_3$ human skin collagen, release by pepsin digestion and preponderance in fetal skin. J Biol Chem 1974; 249:3225-3231.
12. Uitto J, Perejda AJ, Abergel RP, Chu M-L, Ramirez F. Altered steady-state ratio of type I/III procollagen mRNAs correlates with selectively increased type I procollagen biosynthesis in cultured keloid fibroblasts. Proc Natl Acad Sci USA 1985; 82:5935-5939.
13. Uitto J. Biochemistry of the elastic fibers in normal connective tissues and its alterations in disease. J Invest Dermatol 1979; 72:1-10.
14. Uitto J, Paul JL, Brockley K, Pearce RJ, Clark JG. Elastic fibers in human skin: quantitation of elastic fibers by computerized digital image analyses and determination of elastin by a radioimmunoassay of desmosine. Lab Invest 1984; 49:499-505.
15. Starcher BC. Determination of the elastin content of tissue by measuring desmosine and isodesmosine. Anal Biochem 1977; 79:11-15.
16. Gray WR, Sandberg LB, Foster JA. Molecular model for elastin structure and function. Nature 1973; 246:461-466.
17. Comper WD, Laurent TC. Physiological function of connective tissue polysaccharides. Physiological Reviews 1978; 58:255-315.
18. Pearce RH, Grimmer BJ, Mathieson JM. Fractionation of rat cutaneous glycosaminoglycans using an anion-exchange resin. Anal Biochem 1972; 50:63-72.
19. Vogel, KG, Paulsson M, Heinegård D. Specific inhibition of type I and type II collagen fibrilogenesis by the small proteoglycan of tendon. Biochem J 1984; 223:587-597.
20. Uldbjerg N, Danielsen CC. A study of the interaction in vitro between type I collagen and a small dermatan sulphate proteoglycan. Biochem J 1988; 251:643-648.
21. Zimmerman DR, Dours-Zimmermann MT, Schubert M, Bruckner-Tuderman L. Versican is expressed in the proliferating zone in the epidermis and in association with the elastic network of the dermis. J Cell Biol 1994; 124:817-825.
22. Warren R, Gartsetin V, Kligman AM, Montagna W, Allendorf RA, Ridder GM. Age, sunlight, and facial skin: a histologic and quantitative study. J Am Acad Dermatol 1991; 25:751-760.
23. Frances C, Robert L. Elastin and elastic fibers in normal and pathologic skin. Int J Dermatol 1984; 23:166-179.
24. Montagna W, Kirchner S, Carlisle K. Histology of sun-damaged skin. J Am Acad Dermatol 1989; 21:907-918.
25. Kligman LH. Skin changes in photoaging: characteristics, prevention and repair. In: Balin A, Kligman A, eds. Aging and the Skin. New York: Raven Press, 1989:331-346.
26. Taylor CR, Stern RS, Leyden JJ, Gilchrest BA. Photoaging/photodamage and photoprotection. J Am Acad Dermatol 1990; 22:1-15.

27. Kligman AM. Early destructive effects of sunlight on human skin. JAMA 1969; 210:2377-2380.
28. Gilchrest BA. Skin aging and photoaging: an overview. J Am Acad Dermatol 1989; 21:610-613.
29. Chen VL, Fleischmajer R, Schwartz E, Palia M, Timpl R. Immunochemistry of elastotic material in sun-damaged skin. J Invest Dermatol 1986; 87:334-337.
30. Mera SL, Lovell CR, Jones RR, Davies JD: Elastic fibres in normal and sun-damaged skin: an immunohistochemical study. Br J Dermatol 1987; 117:21-27.
31. Bernstein EF, Chen YQ, Tamai K, Shepley KJ, Resnik KS, Zhang H, Tuan R, Mauviel A, Uitto J. Enhanced elastin and fibrillin gene expression in chronically photodamaged skin. J Invest Dermatol 1994; 103:182-186.
32. Dahlback K, Ljungquist A, Lofberg H, Dahlback B, Engvall E, Sakai LY. Fibrillin immunoreactive fibers constitute a unique network in the human dermis: immunohistochemical comparison of the distributions of fibrillin, vitronectin amyloid p component and orcein stainable structures in normal skin and elastosis. J Invest Dermatol 1990; 94:284-291.
33. Bernstein EF, Fisher LW, Li K, LeBeron RG, Tan E, Uitto J. Differential expression of the versican and decorin genes in photoaged and sun-protected skin: Comparison by immunohistochemical and Northern analyses. Lab Invest 1995; 72:662-669.
34. Bernstein EF, Underhill CB, Hahn PJ, Brown DB, Uitto J. Chronic sun-exposure alters both the content and distribution of dermal glycosaminoglycans. Br J Dermatol 1996; 135:255-262.
35. Uitto J, Matsuoka LY, Kornberg RL. Elastic fibers in cutaneous elastoses. In: Rudolph R, ed. Problems in Aesthetic Surgery. St. Louis: CV Mosby Co. 1986:307-338.
36. Griffiths CEM, Russman AN, Majmudar G, Singer RS, Hamilton TA, Voorhees JJ. Restoration of collagen formation in photodamaged human skin by tretinoin (retinoic acid). N Engl J Med 1993; 29:530-535.
37. Schwartz E, Cruickshank FA, Christensen CC, Perlish JS, Lebwohl M. Collagen alterations in chronically sun-damaged human skin. Photochem Photobiol 1993; 58:841-844.
38. Oikarinen A, Kallioinen M. Biochemical and immunohistochemical study of collagen in sun-exposed and protected skin. Photodermatol 1989; 6:24-31.
39. Bernstein EF, Chen YQ, Kopp JB, Fisher L, Brown D, Hahn P, Robey FA, Uitto J. Chronic sun-exposure alters the collagen of the superficial dermis: Evaluation by Northern analysis, immunohistochemical staining and confocal scanning laser microscopy. J Am Acad Dermatol 1996; 34:209-218.
40. Bissett DL, Hannon DP, Orr TX. An animal model of solar-aged skin: histological, physical and visible changes in UV-irradiated hairless mouse skin. Photochem Photobiol 1987; 46:367-376.

41. Zimmerman DR, Dours-Zimmermann MT, Schubert M, Bruckner-Tuderman L. Versican is expressed in the proliferating zone in the epidermis and in association with the elastic network of the dermis. J Cell Biol 1994; 124:817-25.
42. Bianco P, Fisher LW, Young MF, Termine JD, Robey PG. Expression and localization of the two small proteoglycans biglycan and decorin in developing human skeletal and nonskeletal tissues. J Histochem Cytochem 1990; 38:1549-63.
43. Fleischmajer R, Fisher LW, MacDonald ED, Jacobs L Jr., Perlish JS, Termine JD. Decorin interacts with fibrillar collagen of embryonic and adult human skin. J Struct Biol 1991; 106:82-90.
44. Vogel KG, Paulsson M, Heinegård D. Specific inhibition of type I and type II collagen fibrillogenesis by the small proteoglycan of tendon. Biochem J 1984; 223:587-97.
45. Uldbjerg N, Danielsen CC. A study of the interaction in vitro between type I collagen and a small dermatan sulphate proteoglycan. Biochem J 1988; 251:643-8.
46. Scott JE, Orford CR. Dermatan sulphate-rich proteoglycan associates with rat tail-tendon collagen at the d band in the gap region. Biochem J 1981; 197:213-6.
47. Hsu-Wong S, Katchman SD, Ledo I, Wu M, Khillan J, Bashir M, Rosenbloom J, Uitto J. Tissue-specific and developmentally regulated expression of human elastin promoter activity in transgenic mice. J. Biol. Chem. 1994; 269:18072-18075.
48. Bernstein EF, Brown DB, Urbach F, Forbes D, Del Monaco M, Wu M, Katchman SD, Uitto J. Ultraviolet radiation activates the human elastin promoter in transgenic mice: A novel in vivo and in vitro model of cutaneous photoaging. J Invest Dermatol 1995; 105:269-273.
49. Bernstein EF, Brown DB, Takeuchi T, Kong SK, Uitto J. Evaluation of sunscreens with various sun protection factors in a new transgenic mouse model of cutaneous photoaging. J Am Acad Dermatol-in press
50. Bernstein EF, Gasparro FP, Brown DB, Tsunemichi T, Uitto J. 8-methoxypsoralen and ultraviolet A radiation activate the human elastin promoter in transgenic mice: In vivo and in vitro evidence for gene induction. Photochem Photobiol 1996; 64:369-374.
51. Tsunemichi T, Uitto U, Bernstein EF. A novel in vivo model for evaluating agents which protect against ultraviolet A-induced photoaging. Arch Dermatol (submitted for publication).

CHAPTER 7

Immune Aspects of Sunscreens

Peter Wolf and Margaret L. Kripke

Introduction

Exposing the skin to ultraviolet (UV) radiation induces many biologic alterations, including a variety of effects on the immune system (reviewed in refs. 1,2). For example, UV radiation suppresses cell-mediated immune responses and can lead to a state of long-term, antigen-specific, immune suppression. In mice, UV-induced immune suppression contributes to the development of skin cancer. There is evidence that UV-induced alterations in immune function also play a significant role in skin carcinogenesis in humans. In addition, UV-induced immune suppression increases the severity of infectious diseases in animal models,[1,3] and may contribute to the pathogenesis of some infectious diseases in humans. Sunscreens are designed primarily to prevent the acute effects of UV radiation.[4] Their efficacy is indicated by a sun protection factor (SPF), which is based on their ability to prevent erythema in human skin. The quest for better sunscreens resulted in significant improvements, leading to preparations with SPFs of 40 or more. These preparations are highly protective against the acute effect of UV radiation, i.e., the sunburn reaction. In animal studies, sunscreens also protected against chronic UV-induced skin aging,[5] tumor initiation,[6,7] and tumor promotion.[8] However, the ability of sunscreens to protect against the immunosuppressive effects of UV radiation has been the subject of much controversy (reviewed in refs. 9, 10). Insufficient protection from immune suppression by sunscreens may pose a risk with regard to skin cancer formation, particularly, when sunscreens are used to prolong sun exposure.

Sunscreen Photobiology: Molecular, Cellular and Physiological Aspects,
edited by Francis P. Gasparro.

UV-Induced Immune Suppression and UV Carcinogenesis

UV radiation is the major carcinogen in the etiology of basal and squamous cell carcinomas in humans,[11] and induces a variety of skin cancers in experimental animals.[6] In mice, immune suppression by UV radiation plays an important role in the pathogenesis of these skin cancers.[12] Exposure of mice to UV radiation suppresses their ability to reject transplanted UV-induced tumors, which are highly antigenic. Ways in which UV radiation can cause immune suppression are by impairing antigen presentation,[13,14] and increasing the production of immunosuppressive cytokines.[15] Both mechanisms contribute to the formation of antigen-specific suppression and immunologic tolerance. In addition, UV irradiation can inhibit cell-mediated immune responses elicited in UV-irradiated skin by unknown mechanisms.[16]

In humans, the possible importance of immune surveillance mechanisms in skin carcinogenesis was suggested by the observation that organ transplant recipients undergoing immunosuppressive therapy have a high incidence of skin cancers on sun-exposed body surfaces.[17,18] The finding of a recent study[19] also strongly supports the hypothesis that UV-induced immune suppression contributes to skin cancer formation in humans. In this study, exposure of skin to low dose UV radiation separated humans into two phenotypically distinct groups, termed UVB-resistant and UVB-susceptible, which were revealed when the contact allergen dinitrochlorobenzene (DNCB) was applied to UV-irradiated skin. UVB-susceptible individuals generated no contact hypersensitivity (CHS) response when exposed to UVB and contact allergen, whereas UVB-resistant individuals developed vigorous CHS. In normal, healthy individuals, the incidence of UVB-susceptibility was within the range of 30-40%. In contrast, 92% of patients with a history of basal or squamous cell carcinomas were characterized as UVB-susceptible,[19] suggesting that susceptibility to UV-induced immune suppression is an added risk factor for skin cancer development.

The genetic disease xeroderma pigmentosum has also provided evidence for the possible role of the immune system in the development of skin cancers.[20-24] Patients with xeroderma pigmentosum are hypersensitive to sunlight, prone to develop multiple skin cancers, and defective in cellular repair of UV-induced DNA damage. Since cells of all organs presumably share the DNA repair defect, it was postulated that these patients should show an exaggeration of any immunosuppressive effect of UV radiation mediated by DNA damage and that such immune suppression may play a role in their susceptibility to skin cancer.[1] Indeed, various defects of cell-mediated immunity have been reported in patients with xeroderma pigmentosum, including impaired cutaneous responses to DNCB[20] and recall antigens,[20,21] decrease in the ratio of circulating T helper/suppressor cells,[21] impaired lymphocyte proliferative responses to phytohemagglutinin,[20,21] and defective natural killer cell activity.[22] In one study,[24] the degree of impairment of DNCB sensitization correlated directly with the severity of the cutaneous disease. In contrast, patients with Cockayne's syndrome and trichothiodystrophy, two other genetic diseases with deficient DNA repair mechanisms, do not exhibit immunological alterations,[22] nor do they have an increased incidence of skin cancer. The hypothesis that immune suppression may contribute to skin cancer formation in xeroderma pigmentosum is also supported by an unfortunate study in which patients with xeroderma pigmentosum were treated with PUVA therapy in an attempt to induce a protective tan.[23] A sudden onset of multiple tumors was observed in the xeroderma pigmentosum patients within a few weeks after starting PUVA treatment. Because of the rapid onset of the tumors, and because PUVA treatment has effects on the immune system similar to those of UVB radiation,[2] it was suggested that PUVA-induced suppression of immune surveillance mechanisms allowed the outgrowth of pre-existing transformed cells.[23]

There is also epidemiologic evidence that UV radiation is involved in the etiology of cutaneous melanoma.[25] However, in contrast to cutaneous carcinomas, which are primarily related to chronic, cumulative UV exposure[11,26] the role of UV radiation in the development of cutaneous melanoma is not defined.[25] Evidence from animal studies suggests that UV-induced immunologic alterations may also play a role in the development of cutaneous melanoma.[27,28] Mice exposed to UV irradiation and injected with syngeneic melanoma cells in the external ear exhibit an increased incidence of melanoma compared to unirradiated mice.[27] That this effect of UV radiation on melanoma growth is immunologically mediated was supported by the observation that it was only found for highly immunogenic tumors and not for nonimmunogenic tumors.[28] Moreover, this UV effect on melanoma growth was not observed in mice already immunosuppressed by congenital athymia, sublethal X-irradiation, or antibody-mediated T cell depletion in vivo.[28] It is difficult to extrapolate these findings to humans; however, there is evidence that immunologic factors are important in the development of cutaneous melanoma in humans.[29,30] For instance, the occasional regression of cutaneous melanoma may be caused by a lymphocytic tumor infiltrate with an increased ratio of CD4/CD8-positive cells,[31] and specific immunotherapy of cutaneous melanoma can be successful.[29,30]

Recently, it has been suggested that UV-induced immune suppression may play a role in the increasing incidence of non-Hodgkin's lymphoma.[32] The incidence of this lymphoid malignancy exhibits a similar latitude gradient to squamous cell carcinoma, and its increase in recent years parallels that of skin cancer. Epidemiologic studies demonstrating that persons who have had non-Hodgkin's lymphoma have an increased risk of skin cancer development, and vice versa, strengthen the association between non-Hodgkin's lymphoma and exposure to UV radiation.[33] Because immune suppression is a predisposing factor in this disease, there is speculation that immune suppression resulting from exposure to UV radiation may account for the unexpected geographic distribution of non-Hodgkin's lymphoma and its apparent association with solar radiation.

UV Radiation and Infectious Diseases

It has long been known that sunlight can trigger cold sores on the lips in persons harboring latent infection with *Herpes simplex* virus type 1 (HSV-1), and it was demonstrated that this induction is UV dose dependent.[34] Similar results were found with herpes genitalis, presumably due to *Herpes simplex* virus type 2 (HSV-2).[35] Local alterations of immunity may be responsible for reactivation of the latent herpes simplex infection, but there is no exact information on the mechanism.[1] In mice, UV irradiation at the site of intradermal infection with HSV-2 increased the severity of the disease, and there was systemic suppression of delayed type hypersensitivity (DTH) to the virus and generation of antigen specific suppressor T lymphocytes.[36] Similar results were found in a model of HSV-1 infections in mice.[37] (See ref. 38 for a review on mouse models of herpes simplex virus infection.)

Exposure of mice to UV radiation also caused systemic suppression of DTH to the opportunistic yeast, *Candida albicans*.[39,40] In humans, *C. albicans* is frequently responsible for minor infections of the skin and mucous membranes in normal subjects; however, life-threatening infections by *C. albicans* can occur in immunosuppressed persons. In addition, UV irradiation of mice decreased the DTH response and dramatically altered the course of infection with *Mycobacterium bovis* bacillus Calmette-Guérin (BCG), an organism closely related to the bacterium that causes tuberculosis in humans.[41] After experimental infection, many more viable bacilli were found in the peripheral lymph nodes of UV-irradiated mice than in unirradiated mice.[41] Similar results were observed with *M. lepraemurium*, a pathogen that causes a chronic, lethal infection in mice. In an experimental study in humans, repeated local exposure to UV radiation downregulated the elicitation of a cellular immune response to leprosy antigens and decreased the formation of lepromin-induced granulomas in sensitized subjects at the site of UV exposure.[42] Studies with the protozoan parasite *Leishmania major* demonstrated that experimental infection of mice through UV-irradiated skin also impaired their DTH response to Leishmania antigens and reduced their resistance to reinfection.[43,44] Exposure to UV radiation also modulated the immunity of mice to *Borrelia burgdorferi* and led to differential effects on Th-1 and Th-2 type immune responses.[45]

Alterations of the immune response by UV radiation may be particularly important in persons who are human immunodeficiency virus (HIV) positive. There is concern that exposure of HIV^+ subjects to UV radiation may play a role in the development and/or progression of the acquired immunodeficiency syndrome (AIDS).[46-48] This concern is based on demonstrations in vitro and in transgenic mice that UV irradiation can activate the HIV promoter.[49] In addition, UV exposure could also exacerbate progression of AIDS by interfering with protective immunity. For instance, UV-induced injury of Langerhans cells which harbor HIV may result in impaired immunity to HIV due to decreased antigen presentation.[50] The UV-induced increase of interleukin (IL)-10 production in HIV^+ subjects may reduce the potential immunoprotective effect provided by T helper 1-type responses that augment cellular immunity.[51,52] On the other hand, phototherapy has been successfully used in the treatment of various skin diseases in HIV^+ patients.[53-55] In one study,[55] UVB phototherapy did not significantly alter the systemic immune status of AIDS patients and half-body exposure improved the cutaneous symptoms only within the treated areas and reduced the total number of T lymphocytes within the skin lesions. Thus, the authors suggested that UVB radiation may not exacerbate the systemic immunosuppressive state already present in AIDS patients, but instead, may only modify local immunity. However, the risk/benefit ratio of UV exposure may vary depending on the stage of HIV disease.[53]

Garssen et al[3] have recently performed a quantitative risk assessment of lowered resistance to infectious diseases in humans due to UV exposure. The authors calculated, based on interspecies comparison of the dose responses for the suppression of lymphocyte proliferation to *Listeria monocytogenes* in rats and humans, that approximately 100 min of solar exposure at around noon in July in Southern Europe would suppress resistance to infection by *L. monocytogenes* in the most sensitive humans.[3]

Mechanisms of UV-Induced Immune Suppression

Mice exposed chronically to UV radiation are unable to reject highly antigenic, UV-induced skin cancers.[12] In adoptive transfer experiments, it was demonstrated that this systemic effect of UV radiation is caused by suppressor T lymphocytes that prevent immune surveillance against UV-induced skin cancers.[56] The mechanisms by which UV radiation impairs immune responses and leads to induction of suppressor T lymphocytes has been studied extensively in animals using the model of UV-induced suppression of the induction of contact hypersensitivity (CHS).[2]

Exposure of rodents to UV radiation suppresses the induction of CHS to haptens when sensitization is performed after UV exposure. The immunologic impairment caused by UV radiation is specific for the hapten employed and can be divided into local and systemic suppression. Local immune suppression is defined as the diminished CHS response observed when haptens are applied through UV-irradiated skin. Exposure to UV radiation also results in a diminished CHS response when haptens are applied at a distant, unirradiated site; this is referred to as systemic immune suppression.[2] Some of the features of UV-induced suppression of CHS resemble those involved in UV-induced susceptibility to tumor growth, namely, altered antigen presentation and induction of T-suppressor cells.[57]

An alteration in the function of epidermal Langerhans cells has been implicated in local immune suppression.[58] The production of soluble factors from cells in the skin may be involved in both local[59] and systemic UV-induced immune suppression.[15,60-63] Various soluble mediators (for a review see ref. 63) such as prostaglandins,[60] tumor necrosis factor-α ,[59] interleukin (IL)-1,[60] contra-IL-1,[61] and IL-10[15,63] have been implicated as possible mediators of immune suppression by UV radiation. In regard to infectious diseases, IL-10 has been recently shown to exhibit a major role in UV-induced suppression of both DTH to *M. bovis* BCG and impairment in the clearance of these bacteria.[64] Furthermore, urocanic acid in the stratum corneum has immunosuppressive properties, particularly after isomerization to its *cis* form by UV irradiation.[59,65] Although DTH and CHS responses are quite similar, different mechanisms may be involved in their suppression by UV radiation. For instance, different soluble immunosuppressive factors with different wavelength dependencies seem to be involved.[62] At the molecular level, suppression of CHS and suppression of DTH to certain antigens seem to be mainly mediated by DNA damage in the form of cyclobutane pyrimidine dimers.[66-68] However, suppression of DTH to other antigens, notably HSV[69] and alloantigen[70] appear to be mediated by isomerization of urocanic acid.

Recently, it was demonstrated that exposure to UV radiation also suppresses the induction of CHS and DTH responses in humans.[18,68] The mechanisms of UV-induced immune suppression in humans may be similar to those in rodents. For instance, similar morphological alterations occur in epidermal Langerhans cells of mouse and human skin exposed to UVB radiation, and both mice and humans have increased levels of *cis*-urocanic acid in their skin and IL-1 activity in their serum after exposure to UVB.[1,58] In UV-susceptible people, sensitization with DNCB through UV-irradiated skin not only decreased the CHS response, but also induced immunological tolerance in some individuals.[19]

Action Spectrum of UV-Induced Immune Suppression

The wavelengths responsible for UV-induced immune suppression are primarily in the UVB (280-320 nm) range.[1,2] However, the exact action spectrum for a particular immunosuppressive effect of UV may depend on the type of immune response.[62] Wavelengths in the UVA (320-400 nm) range alone were not found to be immunosuppressive in mice in vivo, although morphologic alterations of cutaneous immune cells[72] and isomerization of urocanic acid[73] were observed after UVA exposure.

Composition and Concentration of Sunscreens Used for Immune Protection

Two major types of sunscreen agents are available for sunscreen formulation: (1) chemical UV filters, i.e., octyl-N-dimethyl-p-aminobenzoate (o-PABA), 2-ethylhexyl-p-methoxycinnamate (2-EHMC), benzophenone-3, and other agents, which protect by absorption of certain wave bands of UV radiation; and (2) physical sunscreens, i.e., zinc oxide, titanium dioxide, and other pigments, which protect by reflection, scattering, and/or absorption of UV radiation throughout the entire UV spectrum and even in the near visible range, depending on the particle size. In most cases modern sunscreens contain a combination of one or two chemical UVB filters, one or two UVA filters, and additional pigments. The majority of sunscreen studies for immune protection employed preparations containing chemical UV filters alone;[10] however, recently sunscreen preparations containing only pigments as active ingredients were examined also.[74]

The quantity of sunscreen applied to the skin is an important factor in preventing the acute effects of sunlight[75,76] and may also be a significant variable in protection against the immunosuppressive effects of UV radiation. Usually, a concentration of 2 mg or 2 $\mu l/cm^2$ is recommended and used in clinical studies. The use of lower concentrations has been shown to result in a significant loss of photoprotection, at least with regard to sunburn.[75] The importance of the amount of sunscreen for immune protection can be deduced from the results of a study in which immune protection from the suppression of the mixed epidermal cell-lymphocyte reaction (MECLR) was only observed after a second application of the sunscreen 2-EHMC halfway through the irradiation.[77]

Effect of Sunscreens on UV-Induced Immunologic Alterations in Laboratory Animals

Many studies have addressed the question of whether sunscreens can protect laboratory animals from the immunosuppressive effects of UV radiation (reviewed in refs. 9,10). However, the results of these studies vary greatly and have, therefore, led to controversy as to the effectiveness of sunscreens in preventing UV-induced immune suppression. In various studies, the best immunoprotective capacity of sunscreens ranged from no protection,[74,78-85] or partial protection,[77,79,85-88] to complete protection,[69,74,81,84,87,89-92] depending on the experimental protocol and sunscreen being examined. In certain studies sunscreens completely failed to protect rodents against various UV-induced immunological alterations such as local[79,82] and systemic[80] suppression of CHS, induction of susceptibility to tumor transplantation,[78,85] and enhancement of melanoma growth.[83] Importantly, several studies revealed that chemical sunscreens have a good capacity to protect against UV-induced qualitative and quantitative alterations of Langerhans cells, although sometimes the protection was incomplete.[69,79,82,93] However, it was reported that protection from changes in Langerhans cells does not imply protection from alterations in immune function.[69,79,82] Differences in methodology (i.e. UV light source, number of UV treatments, UV dose, and sunscreen formulation, concentration, and mode of application) may have accounted for some of the discrepancies in protection of immune function among the different studies. Most importantly, UV dose-responses were not performed in the earlier studies on immune protection.[78-82,85,88,91] If the total UV dose exceeded the protective capacity of a sunscreen, no effect of the sunscreen would have been observed, even though it may have been immunoprotective at lower UV doses. Thus, high UV doses and lack of dose-response information might explain why immune protection by sunscreens was not noted in some earlier studies.[78-82,85]

More recently, several investigators compared in dose response studies the capacity of sunscreens to protect from immune suppression to their capacity to protect from inflammation and introduced the concept of immune protection factors (Table 7.1).[74,83,87,89,90,92] Wolf et al compared the photoprotective properties of sunscreen preparations containing the chemical UV filters 2-EHMC, o-PABA, or benzophenone-3 in four separate studies using unfiltered FS40 sunlamps in C3H mice.[68,83,86,87] In the first study, using the well characterized model of UV-induced local and systemic suppression of CHS to 2,4-dinitrofluorobenzene (DNFB), all three sunscreens gave complete protection against sunburn and local suppression of CHS caused by a single dose of 200 mJ/cm^2 UVB.[87] This dose represented approximately 2 minimal edema doses in the mice. The effect of the sunscreens on systemic suppression of CHS was studied across a broad range of UVB doses from 1 to 16 minimal edema doses. Protection was highly dependent on the UV dose, in that complete protection was observed at low doses of UVB, but not at higher doses. The sunscreens were less effective in protecting against systemic suppression

Table 7.1. Studies directly comparing the immune protective and anti-inflammatory capacity of sunscreens in mice

Reference	UVR source	Dosage (mJ/cm^2)	Species	Inflammatory response
Wolf et al[87]	FS40 sunlamps	1 x 200-3200	C3H/HeN (MTV^-) mice	Skin swelling[c]
Wolf et al[83]	FS40 sunlamps	6 x 480	C3H/HeN mice	Skin swelling[c]
Roberts and Beasley[89]	FS20 sunlamps, KF FS20 sunlamps or xenon arc solar simulator	2 x 25-8100	C3H/HeNHsd mice	n.d.
Bestak et al[90]	KF Cosmolux RA plusA1 UVA + FS72T12-UVB-HO tubes (CHS) / xenon arc lamp (skin swelling)	20 x 240-527	Balb/c mice	Skin swelling[c]
Roberts et al[90]	FS20 sunlamps, KF FS20 sunlamps or xenon arc solar simulator	2 x 25-16200	C3H/HeNHsd mice	n.d.
Walker and Young[92]	Philips T101 fluorescent tubes	1 x 250-6181	Hairless HRA.HRII-c/+/SKH mice	Skin swelling[c]

[a] SPF, sun protection factor, based on protection from inflammation
[b] IPF, immune protection factor, based on protection from the suppression of a specific immune response
[c] Skin swelling as measured by mid dorsal double-skinfold thickness
[d] labeled human SPF
[e] The IPF varied depending on the light source used for testing; IPF for xenon arc solar simulator > Kodacel filtered FS20 sunlamps > unfiltered FS20 sunlamps

Table 7.1 (cont.)

Immune repsonse	Sunscreen (concentration)	SPF[a]	IPF[b]
Systemic suppression of CHS to DNFB	o-PABA (7.5)	≥8	≤4
	2-EHMC (8)	≥8	≤4
	BP-3 (6)	≥4	≤2
Growth enhancement of transplanted	o-PABA (7.5)	≥8	≤1
	2-EHMC (8)	≥8	≤1
	BP-3(6)	≥4	≤1
Local suppression of CHS to DNFB	2-EHMC (7.5) + BP-3 (4)	15[d]	2-30[e]
Systemic suppression of CHS to TNCB	o-PABA (8)	4	≤1
	2-EHMC (8)	4	1-1.1
	MTD (7.5)	7	1-1.1
Local suppression of CHS to DNFB	2-EHMC (3.5) + BP-3 (1)	4[d]	1-15[e]
	2-EHMC (7) + BP-3 (2)	8[d]	2-15[e]
	2-EHMC (7.5) + BP-3 (4)	15[d]	4-30[e]
	2-EHMC (7.5) + OS (5) + HS (5) + BP-3 (1)	30[d]	4-60[e]
Systemic suppression of CHS to DNFB	2-EHMC (6.3)	4	4

Abbreviations:
BP-3, benzophenone-3
CHS, contact hypersensitivity
DNFB, 2,4-dinitrofluorobenzene
HS, homosalate
KF, Kodacel filtered (i.e. wavelengths below 290 nm removed)
MTD, microfine titanium dioxide
n.d., not done
OS, octyl salicylate
o-PABA, octyl dimethyl para-aminobenzoic acid
TNCB, trinitrochlorobenzene
UV, ultraviolet radiation
2-EHMC, 2-ethylhexyl-4'-methoxycinnamate

of CHS than against sunburn. In the second study,[86] the sunscreens very efficiently protected against sunburn and cyclobutane pyrimidine dimer formation, but only partially protected against suppression of DTH to *C.albicans* caused by a single dose of 500 mJ/cm^2 UVB radiation. In the third study,[83] Wolf et al investigated the effect of the sunscreens on melanoma outgrowth in a C3H mouse transplantation model, in which enhanced tumor growth is mediated by a local immune mechanism,[28] unrelated to inflammation[94] and local suppression of CHS.[95] In this study, the sunscreens completely lacked the capacity to protect against melanoma outgrowth produced by a minimal effective dose of 480 mJ/cm^2 of UVB radiation given twice per week for 3 weeks, under circumstances in which they completely inhibited the sunburn reaction and clearly reduced microscopic alterations, including sunburn cell formation.[83] In the fourth study,[68] Wolf et al reported that their sunscreen preparations completely protected C3H mice from UV-induced local suppression of CHS to DNFB and nearly completely protected from sunburn cell formation and quantitative and qualitative changes of cutaneous immune cells, including Thy-1$^+$ dendritic epidermal T cells and Langerhans cells, produced by a single dose of 500 mJ/cm^2 UVB radiation from FS40 sunlamps. Taken together, the results of these studies clearly indicated that at a given UV dose (i.e. approximately 500 mJ/cm^2 UVB radiation) the immunoprotective capacity of the sunscreens ranged from no protection to complete protection, depending on the type of immune response examined;[68,83,86,87] Immune protection factors could be determined from two studies;[83,87] they ranged between ≤ 1 and ≤ 4 for the UVB filters o-PABA and 2-EHMC, and ≤ 1 and ≤ 2 for the UVA filter benzophenone-3, whereas the conventional SPF of the sunscreens was ≥ 8 and ≥ 4, respectively (Table 7.1).

Bestak et al[74] recently compared the ability of the chemical UV filters 2-EHMC and o-PABA, and the physical sunscreens microfine titanium dioxide and zinc oxide to protect the immune system of Balb/c and C3H mice from chronic solar-simulated UV irradiation over 4 weeks. Using a suberythemal protocol that induced local but not systemic immune suppression or tolerance to TNCB in Balb/c mice the o-PABA-containing preparation increased the immune suppression, whereas the 2-EHMC- and microfine titanium dioxide-containing preparation protected the immune system. When the cumulative dose was increased by 12.7%, causing systemic immune suppression, none of the sunscreens protected from systemic suppression of the induction of CHS, but 2-EHMC provided partial, and microfine titanium dioxide, complete protection from tolerance. In C3H mice, which had an MED similar to that of Balb/c mice but were systemically immunosuppressed with only 44% of the UV dose required for Balb/c mice, 2-EHMC gave some protection, whereas microfine titanium dioxide and zinc oxide gave complete protection from systemic immune suppression. The authors concluded that their sunscreens gave limited immune protection, which was not related to the SPF of the sunscreens or the MED of the mouse strain, but rather was linked to the sunscreens' having a broad absorption spectrum extending into the UVA range. They calculated a very low immune protection factor for systemic suppression of CHS of 1.13 for both the 2-EHMC- and the microfine titanium dioxide-preparation, whereas the conventional SPF of these sunscreens for the mice was 4.0 and 7.0, respectively (Table 7.1). There was some correlation between the UV effects on the immune system and histologic alterations. The o-PABA containing preparation, which was the least immunoprotective sunscreen in both mouse strains, actually induced epidermal thickening, keratinization, and dermal changes by itself in both Balb/c and C3H/HeJ mice. Importantly, in C3H mice, o-PABA, 2-EHMC, and microfine titanium dioxide suppressed CHS in the absence of UV irradiation. In order to examine the effects of the sunscreens in the absence of immune suppression induced by the sunscreens themselves, the investigators added the oxygen radical scavengers butylated hydroxytoluene and α-tocopherol to the base lotion. This observation clearly indicated that sunscreens themselves can interact with the skin, and, possibly via an oxygen-dependent mechanism, can lead to changes in the immune response. This hypothesis is supported by recent findings in which singlet molecular oxygen was formed inlluminated solutions of several commercially available, active sunscreen ingredients.[97]

Walker and Young[92] recently used monochromatic UVB radiation to remove possible confounding effects of differences in action spectra between inflammation and immune suppression, the endpoint examined for photoprotection. The authors reported that sunscreen preparations containing o-PABA or 2-EHMC only partially protected hairless HRA.HRII-c/ + /Skh mice from systemic suppression of CHS to DNFB by narrow-band 311 nm UVB irradiation, under conditions in which the sunscreens completely protected against skin edema. The results were the same for the application of the sunscreens directly to the skin or to transpore tape above the irradiation cages. Dose-response studies for edema and systemic suppression of CHS revealed that suppression of CHS was more sensitive to UVB irradiation by a factor of 2. Importantly, the authors found an identical protection factor of 4 for 2-EHMC for both systemic immune suppression and inflammation (Table 7.1). Thus, the authors concluded that incomplete immune protection was due in their study to the lower threshold dose for immune suppression. In addition, the identical results after in situ and transpore tape application suggested that the interaction of sunscreens with the skin did not influence immune protection, at least, when monochromatic UV radiation was employed.

In contrast, Roberts and coworkers[89,90] reported, based on studies of local suppression of CHS in C3H mice, that the immunoprotective capacity of several commercial sunscreen preparation greatly exceeded their labeled human SPF, particularly when solar-simulated UV radiation was used (Table 7.1). However, they found poor immunoprotective capacity of the sunscreens when they used filtered or unfiltered FS20 sunlamps for UV irradiation. The authors attributed these results to the fact that FS-type lamps emit a small amount of UVC radiation, which is highly immunosuppressive. Indeed, in a previous study[97] the UVC radiation of FS20 sunlamps contributed 12% and 18%, respectively, of the total edema- and immune suppression-inducing energy. The results of Roberts and coworkers agree with the findings of two previous studies: o-PABA was not effective in preventing the development of tumor susceptibility in mice exposed to FS-type sunlamp bulbs;[85] however, it completely abrogated tumor susceptibility, when natural sunlight was used for UV exposure.[91] Whereas the work by Roberts and coworkers[89,90] clearly indicated that the spectrum of a light source is critical for the evaluation of the immunoprotective capacity of a sunscreen, one problem remains with the interpretation of these studies.[98] In contrast with other investigators,[68,74,83,86,87,92] Roberts and coworkers did not determine the SPF of their sunscreens in murine skin, but rather, compared the immunoprotective capacity of the sunscreens to the labeled SPF for humans. Doing so fails to take into account the possibility that the minimal immunosuppressive dose, the minimal inflammatory dose, and consequently, the immune and inflammatory protection factors, may differ greatly between mice and men.

Effect of Sunscreens on UV-Induced Immunologic Alterations in Humans

Hersey et al[71] first investigated the effect of a sunscreen on immune function in humans (Table 7.2). They reported that an SPF 15 sunscreen containing o-PABA, benzophenone-3, and butyl-methoxy-dibenzoyl-methane failed to protect normal human subjects against decreased natural killer cell activity, as measured against the melanoma cell line MM200 and the K562 myeloid cell line. Suppressed natural killer cell function was produced by 12 exposures to solarium lamps, which emitted both UVA and UVB radiation. In addition, application of the sunscreen also failed to protect against decreased DTH responses to recall antigens and immunoglobulin production by pokeweed mitogen-stimulated cultures of B and T cells in vitro.

Moomaas et al[99] reported that different chemical sunscreens with comparable SPFs gave different degrees of partial protection from UV-induced suppression of the in vitro allostimulatory activity of human peripheral blood mononuclear cells in a mixed leukocyte reaction (Table 7.2). Van Praag et al[100] found that an SPF 6 and an SPF 15 sunscreen failed to protect from UVB-induced suppression of the ability of human epidermal cells, from skin UV-irradiated over 4 weeks, to stimulate allogeneic T lymphocytes (MECLR) (Table 7.2).

Edwards et al[101] and Elmets et al[102] reported that sunscreens have the capacity to protect against UV-induced damage to cutaneous Langerhans cells in humans (Table 7.2). However, studies on immune function were not performed, and there is not an absolute correlation between the number and morphologic integrity of cutaneous immune cells and their ability to induce CHS.[68,79,82]

In an important study, Whitemore and Morison[103] recently reported that an SPF 29 sunscreen, containing 2-EHMC, oxybenzone, and octyl salicylate, protected human volunteers from local suppression of CHS to DNCB, produced by exposure to three consecutive doses of three MEDs (Table 7.2). LeVee et al[104] presented preliminary data suggesting that a commercial sunscreen with an SPF of 4, containing o-PABA as its only sunscreen agent, provided protection from local suppression of CHS to DNCB induced by solar simulated UV radiation (Table 7.2).

Table 7.2. Summary of studies on the immune protective capacity of sunscreens in humans

Reference	UVR source	Dosage/MED[a]	Sunscreen (SPF)	Immunologic Endpoint	Immune Protection
Edwards et al[101]	FS40 sunlamps	1 x 108 mJ/cm^2 UVB	o-PABA+BP-3 (24.5)	Decrease of Langerhans cell number	Partial
Hersey et al[71]	Rellarium solarium lamps	12 x 185 mJ/cm^2 UVB + 17.82 J/cm^2 UVA / ≈ 1 MED	o-PABA+BP-3+4-TBMM (15)	NK cell number	Partial
				NK cell activity	No
				Decrease of CD4/CD8 ratio	No
				Increase of $CD8^+$ cells	No
				Suppression of DTH	No
				Ig production	No
Mommaas et al[99]	Sylvania F20/T21	1 x 120 mJ/cm^2 UVB	4-TBMM+2-PBSA (10)	Human in vitro MLR	Partial
			4-IPDM+MBC+2-PBSA(10)		Partial
			4-TBMM+2-PBSA (6)		Partial
Van Praag et al[100]	Sylvania F75/85 W UV-21	12 x 30-50 mJ/cm^2 UVB / ≈ 1 MED	4-TBMM++MBC+2-PBSA(6)	Human in vivo MECLR	No
			4-TBMM++MBC+2-PBSA+dimethyl-PABA (15)		No

Elmets et al[102]	FS72 sunlamps	4 x 1.5 MED	2-EHMC+BP-3 (15)	Decrease of Langerhans cell number	Complete
Whitemore and Morison[103]	FS72T12/UVB-HO	3 x 3 MED	2-EHMC+BP-3+ OS (29)	Local suppression of CHS to DNCB	Complete
LeVee et al[104]	xenon arc source	1 x 2 MED	o-PABA (4)	Local suppression of CHS to DNCB	Complete
		1 x 4 MED	o-PABA (4)		Partial
		1 x 8 MED	o-PABA (4)		No

[a] Number of exposures x UV dosage and/or multiples of the MED

Abbreviations: BP-3, benzophenone-3; CHS, contact hypersensitivity; DNCB, dinitrochlorobenzene; MBC, 3-(4-methylbenzylidine)-camphor; MED, minimal erythema dose; MECLR, mixed epidermal cell-lymphocyte reaction; MLR, mixed lymphocyte reaction ; NK, natural killer; o-PABA, octyl dimethyl para-aminobenzoic acid; OS, octyl salicylate; SPF, sun protection factor; UVR, ultraviolet radiation; 2-EHMC, 2-ethylhexyl-4'-methoxycinnamate; 2-PBSA, 2 phenyl-benzimidazole-5-sulfonic acid; 4-IPDM, 4-isopropyldibenzoyl methane; 4-TBMM, 4-tert.butyl-4'-methoxy-dibenzoylmethane

Reasons for Insufficient Immune Protection by Sunscreens

Taken together, the existing literature clearly indicates that sunscreens can exhibit immunoprotective capacity. However, the immune protection afforded by sunscreens can be limited and highly variable, depending on the immunological endpoint and experimental protocol. Sunscreens seem to provide better protection against the local immunosuppressive effects of UV radiation,[68,87,89,90,93,101-104] whereas they exhibit much less protection against the systemic effects of UV on the immune system.[71,74,78,80,81,84-88] At a given UV dose, sunscreens provide better protection against sunburn than against immunologic alterations.[77,86,87,92,99] UV-induced immune suppression can occur in the absence of sunburn that is prevented by sunscreen protection.[77,86,87,92] This can be due to a lower threshold dose and/or to a different, e.g., nonparallel, dose response for the immunologic and inflammation endpoints.[92] However, several studies also showed that at least for certain immunologic endpoints under certain conditions, the immune protection factor of a sunscreen can be inferior to its conventional SPF.[74,83,87] Although the reasons for this discrepancy remain largely unknown, several possible explanations may be considered:

(1) Chemical sunscreens do not act as simple physical barriers against UV radiation, but they penetrate the epidermis and may interact with various structures and cellular elements. There may be a nonuniform distribution of sunscreens in the skin, and the target(s), and thus the action spectra, for immune suppression and inflammation may be different or, at best, overlapping; therefore, there may also be a different threshold for photoprotection. This possibility may explain the finding that topical application of liposomes containing the enzyme T4 endonuclease V, specific for the repair of cyclobutane pyrimidine dimers, completely protected against UV-induced suppression of DTH to *C. albicans* under conditions in which the sunburn reaction was only partially inhibited.[68]

(2) It has been suggested that certain sunscreens may pool beneath the stratum corneum,[80] and consequently, although they may inhibit UV radiation-induced damage to deeper structures of the skin, they might not effectively block the access of UV radiation to a superficial chromophore, such as urocanic acid, which may serve as an immunomodulator in the superficial epidermis.[65] Counter to this theory, sunscreens have been shown to protect against *trans*- to *cis*-photoisomerization of urocanic acid in human *stratum corneum* after light exposure.[105] In one study, differential immune protection by

sunscreen was unrelated to *cis*-urocanic acid formation in hairless mice.[84] However, there is also a possibility that another, yet unidentified superficial chromophore, unrelated to urocanic acid may be involved, at least in certain forms of immune suppression.

(3) Several studies in which sunscreens lacked immune protective activity used chronic,[74,77,81-83,85,86,103] rather than single UV exposure regimens. Particularly after repeated UV exposures, certain epidermal and dermal changes, i.e., epidermal hyperplasia and dermal mixed cellular infiltration, may occur despite sunscreen treatment and play a role in UV-induced immune suppression.[83]

(4) Small amounts of DNA damage may be sufficient to cause immunosuppressive effects, despite complete protection from inflammation. This hypothesis is consistent with the findings of a study in which the number of sunburn cells, an indicator of DNA damage,[68,106] although clearly reduced, was not completely prevented by sunscreens, despite complete sunburn protection.[83] The hypothesis is further supported by a study showing that less DNA damage is necessary to trigger immune suppression than inflammation.[86] In contrast, this hypothesis seems to be difficult to reconcile with the findings that sunscreens very effectively prevent DNA damage, particularly cyclobutane pyrimidine dimer formation,[107-110] and that cyclobutane pyrimidine dimers are the main trigger at least of certain forms of immune suppression, i.e., local and systemic suppression of CHS and systemic suppression of DTH to *C. albicans*.[66-68] However, as mentioned before, some types of immune responses may not be triggered solely by formation of cyclobutane pyrimidine dimers.

(5) Protection from formation of cyclobutane pyrimidine dimers in specific regions of DNA, e.g., from actively transcribed regions, rather than overall protection from DNA damage may be critical for immune protection by sunscreens. This hypothesis is indirectly supported by the findings of a study[68] in which two chemical UVB sunscreens gave only partial protection from systemic immune suppression despite preventing detectable cyclobutane pyrimidine dimer formation. However, at the same UV dose, liposomes containing the DNA repair enzyme T4 endonuclease V gave much better immune protection, despite incomplete protection from cyclobutane pyrimidine dimer formation.

(6) Sunscreens may interact chemically with molecules critical for immune suppression, including DNA, and consequently, enhance immune suppression. In support of this theory, increased

mutagenesis was observed in UV-exposed bacteria in the presence of PABA[111,112] and cinnamic acid derivatives[113] related to 2-EHMC. The authors concluded that the sunscreens inhibited the repair of UV-induced DNA damage.[111,113] In addition, PABA augmented the formation of cyclobutane pyrimidine dimers in vitro after UV irradiation.[114,115] In another study,[116] PABA esters alone were able to induce DNA damage, and the damage increased after UV irradiation. There may be a balance between photosensitizing and photoprotective effects of sunscreens: sunscreens might, on the one hand, be able to inhibit immune suppression by preventing direct DNA damage, on the other, they may contribute to immune suppression by inhibiting DNA repair or by augmenting formation of cyclobutane pyrimidine dimers or other DNA photoproducts.

(7) Other or additional mechanisms unrelated to DNA damage may play a role in UV-induced inflammation. For instance, cell membrane damage,[117] as well as DNA damage may be important in UV-induced inflammation. Sunscreens may block such mechanisms more efficiently than formation of DNA damage and in this way provide better protection from inflammation than from immune suppression.

Future Aspects of Sunscreens and Immune Protection

At present, there is a general quest for better, broad spectrum sunscreens with UVA-absorbing properties, because UVA radiation contributes to skin aging[118] and is also carcinogenic.[119,120] The use of UVB-blocking sunscreens alone may allow individuals to receive enormous doses of UVA radiation without the painful sunburn characteristic of UVB exposure, but with other, less obvious negative consequences. From the photoimmunological point of view, UVA-blocking sunscreens may be desirable to prevent photodermatoses such as solar urticaria, polymorphic light eruption, and photosensitive eczema, which are often triggered by wavelengths in the UVA region. Although UVA radiation alone has not been shown to be immunosuppressive in vivo, most studies on immune suppression by UVA were performed in laboratory animals; ultimately, the ability of sunscreens to protect against environmental UV must be assessed in humans exposed to solar UV, using biologically relevant immunologic endpoints.[71,103,104] Better understanding of the mechanisms and wavelengths involved in UV-induced immune suppression should lead to improved sunscreen formulation and broad spectrum photoprotection, including prevention of immune suppression. For instance, in addition to preventing DNA and oxidative damage by chemical and physical sunscreens, DNA repair enzymes or specific antagonists of epidermal cytokines may improve the immunoprotective ability of the next generation of sunscreens.

References

1. Morison WL. Effects of ultraviolet radiation on the immune system in humans. Photochem Photobiol 1989; 50:515-524.
2. Granstein RD. Photoimmunology. Semin Dermatol 1990; 9:16-24.
3. Garssen J, Goettsch W, De Gruijl F et al. Risk assessment of UVB effects on resistance to infectious diseases. Photochem Photobiol 1996; 64:269-274.
4. Pathak MA, Fitzpatrick TB, Frenk E. Evaluation of topical agents that prevent sunburn—superiority of para-aminobenzoic acid and its ester in ethyl alcohol. N Engl J Med 1969; 280:1459-1463.
5. Kligman LH, Akin FJ, Kligman AM. Prevention of ultraviolet damage to the dermis of hairless mice by sunscreens. J Invest Dermatol 1982; 78:181-189.
6. Kligman LH, Akin FJ, Kligman AM. Sunscreens prevent ultraviolet photocarcinogenesis. J Am Acad Dermatol 1980; 3:30-35.
7. Wulf HC, Poulsen T, Brodthagen H et al. Sunscreens for delay of ultraviolet induction of skin tumors. J Am Acad Dermatol 1982; 7:194-202.
8. Synder DS, May M. Ability of PABA to protect mammalian skin from ultraviolet light-induced skin tumors and actinic damage. J Invest Dermatol 1975; 65:543-546.
9. Granstein RD. Evidence that sunscreens prevent UV radiation-induced immunosuppression in humans. Arch Dermatol 1995; 131:1201-1204.
10. Young AR, Walker SL. Photoprotection from UVR-induced immunosuppression. In: Krutmann J, Elmets CA, eds. Photoimmunology. London: Blackwell Science Ltd, 1995: 285-294.
11. Vitaliano PP, Urbach F. The relative importance of risk factors in nonmelanoma carcinoma. Arch Dermatol 1980; 116:454-456.
12. Kripke ML, Fisher MS. Immunologic parameters of ultraviolet carcinogenesis. J Natl Cancer Inst 1976; 57:211-215.
13. Streilein JW, Taylor JR, Vincek V et al. Immune surveillance and sunlight-induced skin cancer. Immunol Today 1994; 15:174-179.
14. Saijo S, Bucana CD, Ramirez KM et al. Deficient antigen presentation and Ts induction are separate effects of ultraviolet irradiation. Cell Immunol 1995; 164:189-202.
15. Rivas JM, Ullrich SE. Systemic suppression of delayed-type hypersensitivity by supernatants from UV-irradiated keratinocytes: an essential role for keratinocyte-derived interleukin-10. J Immunol 1992; 149:3865-3871.

16. Donawho CK, Muller HK, Bucana CD, and Kripke ML. Enhanced growth of murine melanoma in ultraviolet-irradiated skin is associated with local inhibition of immune effector mechanisms. J. Immunol 1996; 157:781-786.
17. Kinlen LF, Sheil AGR, Peto J et al. Collaborative United Kingdom-Australiasian study of cancer in patients treated with immunosuppressive drugs. Br Med J 1979; II:1461-1466.
18. Boyle J, MacKie RM, Briggs JD et al. Cancer, warts, and sunshine in renal transplant patients: a case-control study. Lancet I 1984:702-705.
19. Yoshikawa T, Rae V, Bruins-Slot W et al. Susceptibility to effects of UVB radiation on induction of contact hypersensitivity as a risk factor for skin cancer in humans. J Invest Dermatol 1990; 95:530-536.
20. Dupuy JM, Lafforet D. A defect of cellular immunity in xeroderma pigmentosum. Clin Immunol Immunopathol 1974; 3:52-58.
21. Wysenbeek AJ, Weiss H, Duczyminer-Kahana M et al. Immunologic alterations in xeroderma pigmentosum patients. Cancer 1986; 58:219-221.
22. Norris PG, Limb GA, Hamblin AS et al. Immune function, mutant frequency, and cancer risk in the DNA repair defective genodermatoses xeroderma pigmentosum, Cockayne's syndrome, and trichothiodystrophy. J Invest Dermatol 1990; 4:94-100.
23. Lehmann AR, Bridges BA. Sunlight-induced cancer: some new aspects and implications of the xeroderma pigmentosum model. Br J Dermatol 1990; 122:115-119.
24. Morison WL, Bucana C, Hashem N et al. Impaired immune functions in patients with xeroderma pigmentosum. Cancer Research 1985; 45:3929-3931.
25. Donawho C, Wolf P. Sunburn, sunscreen, and melanoma. Current Opinion in Oncology 1996; 8:159-166.
26. Scotto J, Kopf AW, Urbach F. Nonmelanoma skin cancer in four areas of the U.S. Cancer 1974; 34:1333-1338.
27. Romerdahl CA, Donawho C, Fidler IJ et al. Effect of ultraviolet-B on the in vivo growth of murine melanoma cells. Cancer Res 1988; 48:4007-4010.
28. Donawho CK, Kripke ML. Evidence that the local effect of ultraviolet radiation on the growth of murine melanomas is immunologically mediated. Cancer Res 1991; 51:4176-4181.
29. Mukheryi B, Chakraborty NG. Immunobiology and immunotherapy of melanoma. Curr Opin Oncol 1995; 7:175-184.
30. Donawho CK, Kripke ML. Immunologic factors in melanoma. Clin Dermatol 1992; 10:69-74.
31. Tefany FJ, Barnetson RS, Halliday GM et al. Immunocytochemical analysis of the cellular infiltrate in primary regressing and nonregressing malignant melanoma. J Invest Dermatol 1991; 97:197-202.
32. McMichael AJ, Giles GG. Have increases in solar ultraviolet exposure contributed to the rise in incidence of non-Hodgkin's lymphoma? Br J Cancer 1996; 73:945-950.

33. Levi F, Randimbison L, Te V-C, La Vecchia C. Non-Hodgkin's lymphomas, chronic lymphocytic leukaemias and skin cancers. Br J Cancer 1996; 74(11):1847-1850.
34. Spruance S. Pathogenesis of Herpes simplex labialis: experimental induction of lesions with UV light. J Clin Microbiol 1985; 22:366-368.
35. Wheeler CE. Pathogenesis of recurrent herpes simplex infection. J Invest Dermatol 1975; 65:341-346.
36. Yasumoto S, Hayashi Y, Aurelian L. Immunity to herpes simplex virus type 2: suppression of virus induced immune responses in ultraviolet B irradiated mice. J Immunol 1987; 139:2788-2793.
37. Howie SEM, Norval M, Maingay J. Exposure to low dose UVB light suppresses delayed type hypersensitivity to herpes simplex virus in mice by suppressor cell induction. J Invest Dermatol 1986; 86:125-128.
38. Norval M, El-Ghor AA. UV radiation and mouse models of herpes simplex virus infection. Photochem Photobiol 1996; 64:242-245.
39. Denkins YM, Kripke ML. Effect of UV irradiation on lethal infection of mice with *Candida albicans*. Photochem Photobiol 1993; 57:266-271.
40. Denkins Y, Fidler IJ, Kripke ML. Exposure of mice to UVB radiation suppresses delayed hypersensitivity to *Candida albicans*. Photochem Photobiol 1989; 49:615-619.
41. Jeevan A, Kripke ML. Effect of a single exposure to ultraviolet radiation on *Mycobacterium bovis* Bacillus Calmette-Guerin infection in mice. J Immunol 1989; 143:2837-2843.
42. Cestari TF, Kripke ML, Baptista PL et al. Ultraviolet radiation decreases the granulomatous response to lepromin in humans. J Invest Dermatol 1995; 105:8-13.
43. Giannini SH. Suppression of pathogenesis in cutaneous Leishmaniasis by UV irradiation. Infect Immun 1986; 51:838-843.
44. Giannini SH. Effects of Ultraviolet B irradiation on cutaneous Leishmaniasis. Parasitology Today 1992; 8:44-48.
45. Brown EL, Rivas JM, Ullrich SE et al. Modulation of the immunity to Borrelia burgdorferi by ultraviolet irradiation: differential effect on Th-1- and Th2 type immune responses. Eur J Immunol 1995; 25:3017-3022.
46. Duvic M, Ulmer R, Crane M et al. Treatment of HIV+ patients with UVB is associated with a significant increase in p24 antigen levels. J Invest Dermatol 1995; 104:581 (abstr.).
47. Adams ML, Houpt KR, Cruz PDJ. Is phototherapy safe for HIV-infected individuals? Photochem Photobiol 1996; 64:234-237.
48. Zmudzka BZ, Miller SA, Jacobs ME et al. Medical UV exposures and HIV activation. Photochem Photobiol 1996; 64:246-253.
49. Zmudzka BZ, Beer JZ. Activation of human immunodeficiency virus by ultraviolet radiation. Photochem Photobiol 1990; 52:1153-1162.
50. Tschachler E, Groh E, Popovic M et al. Epidermal Langerhans cells—a target for HTLV-III/LAV infection. J Invest Dermatol 1987; 88:233-237.

51. Clerici M, Shearer GM. UV light exposure and HIV replication. Science 1992; 258:1070-1071.
52. Ullrich SE. Does exposure to UV radiation induce a shift to a Th-2-like immune reaction? Photochem Photobiol 1996; 64:254-258.
53. Smith KJ, Skelton HG, Yeager J. Ultraviolet radiation therapy and HIV disease. J Am Acad Dermatol 1995; 33:841-842.
54. Buchness MR, Lim HW, Hatcher VA et al. Eosinophilic pustular folliculitis in the acquired immunodeficiency syndrome: treatment with ultraviolet B phototherapy. N Engl J Med 1987; 318:1183-1186.
55. Pardo RJ, Bogaert MA, Penneys NS et al. UVB phototherapy of the pruritic papular eruption of the acquired immunodeficiency syndrome. J Am Acad Dermatol 1992; 26:423-428.
56. Fisher MS, Kripke ML. Suppressor T lymphocytes control the development of primary skin cancers in UV-irradiated mice. Science 1982; 216:1133-1134.
57. Noonan FP, De Fabo EC, Kripke ML. Suppression of contact hypersensitivity and its relationship to UV-induced suppression of tumor immunity. Photochem Photobiol 1981; 34:683-689.
58. Stingl G, Gazze-Stingl LA, Aberer W et al. Antigen presentation by murine epidermal Langerhans cells and its alterations by UVB light. J Immunol 1981; 127:1707-1713.
59. Kurimoto I, Streilein JW. Cis-urocanic acid suppression of contact hypersensitivity induction is mediated via tumor necrosis factor-α. J Immunol 1992; 148:3072-3078.
60. Robertson B, Gahring L, Newton R et al. In vivo administration of IL-1 to normal mice decreases their capacity to elicit contact hypersensitivity responses: Prostaglandins are involved in this modification of immune function. J Invest Dermatol 1987; 88:380-387.
61. Schwarz T, Urbanska A, Gschnait F et al. UV-irradiated epidermal cells produce a specific inhibitor of IL-1 activity. J Immunol 1987; 138:1457-1463.
62. Kim T-Y, Kripke ML, Ullrich SE. Immunosuppression by factors released from UV-irradiated epidermal cells: selective effects on the generation of contact and delayed hypersensitivity after exposure to UVA or UVB radiation. J Invest Dermatol 1990; 94:26-32.
63. Ullrich SE. The role of epidermal cytokines in the generation of cutaneous immune reactions and ultraviolet radiation-induced immune suppression. Photochem Photobiol 1995; 62:389-401.
64. Jeevan A, Ullrich SE, De Gracia M et al. Mechanism of UVB-induced suppression of the immune response to *Mycobacterium bovis Bacillus Calmette-Guerin*: role of cytokines on macrophage function. Photochem Photobiol 1996; 64:259-266.
65. De Fabo EC, Noonan FP. Mechanism of immune suppression by ultraviolet irradiation in vivo. I. Evidence for the existence of a unique photoreceptor in skin and its role in photoimmunology. J Exp Med 1983; 157:84-98.

66. Applegate LA, Ley RD, Alcalay J et al. Identification of the molecular target for the suppression of contact hypersensitivity by ultraviolet radiation. J Exp Med 1989; 170:1117-1131.
67. Kripke ML, Cox PA, Alas LG et al. Pyrimidine dimers in DNA initiate systemic immunosuppression in UV-irradiated mice. Proc Natl Acad Sci USA 1992; 89:7516-7520.
68. Wolf P, Cox P, Yarosh DB et al. Sunscreens and T4N5 liposome differ in their ability to protect against UV-induced sunburn cell formation, alterations of dendritic epidermal cells, and local suppression of contact hypersensitivity. J Invest Dermatol 1995; 104:287-292.
69. Norval M, Simpson TJ, Bardshiri et al. Urocanic acid analogues and the suppression of the delayed type hypersensitivity response to herpes simplex virus. Photochem Photobiol 1989; 49:633-639.
70. Moodycliffe AM, Bucana CD, Kripke ML et al. Differential effects of a monoclonal antibody to cis-urocanic acid on the suppression of delayed and contact hypersensitivity following ultraviolet irradiation. J Immunol 1996; 157:2891-2899.
71. Hersey P, MacDonald M, Burns C et al. Analysis of the effect of a sunscreen agent on the suppression of natural killer cell activity induced in human subjects by radiation from solarium lamps. J Invest Dermatol 1987; 88:271-276.
72. Aubin F, Kripke ML. Effect of ultraviolet A radiation on cutaneous immune cells. In: Urbach F, ed. Biological Responses to Ultraviolet A Radiation. Overland Park: Valdenmar Publishing Company, 1992: 239-247.
73. De Fabo EC, Reilly DC, Noonan FP. Mechanism of UVA effects on immune function: preliminary studies. In: Urbach F, ed. Biological Responses to Ultraviolet A Radiation. Overland Park: Valdenmar Publishing Company, 1992: 227-237.
74. Bestak R, Barnetson RSC, Nearn MR et al. Sunscreen protection of contact hypersensitivity responses from chronic solar-simulated ultraviolet irradiation correlates with the absorption spectrum of the sunscreen. J Invest Dermatol 1995; 105:345-351.
75. Stenberg C, Larkö O. Sunscreen application and its importance for the sun protection factor. Arch Dermatol 1985; 121:1400-1402.
76. Bech-Thomsen N, Wulf HG. Sunbathers' application of sunscreen is probably inadequate to obtain the sun protection factor assigned to the preparation. Photodermatol Photoimmunol Photomed 1992-93; 9:242-244.
77. Walker SL, Morris J, Chu AC et al. Relationship between the ability of sunscreens containing 2-ethylhexyl-4'-methoxycinnamate to protect against inflammation, depletion of epidermal Langerhans (Ia^+) cells and suppression of alloactivating capacity of murine skin in vivo. J Photochem Photobiol B Biol 1994; 22:29-36.
78. Gurish MF, Roberts LK, Krueger GG et al. The effects of various sunscreen agents on skin damage and the induction of tumor susceptibility in mice subjected to ultraviolet irradiation. J Invest Dermatol 1981; 76:246-251.

79. Lynch DH, Gurish MF, Daynes RA. Relationship between epidermal Langerhans cell density, ATPase activity and the induction of contact hypersensitivity. J Immunol 1981; 126:1892-1897.
80. Fisher MS, Menter JM, Willis I. Ultraviolet radiation-induced suppression of contact hypersensitivity in relation to Padimate O and oxybenzone. J Invest Dermatol 1989; 92:337-341.
81. Reeve VE, Bosnic M, Boehm-Wilcox C et al. Differential protection by two sunscreens from UV radiation-induced immunosuppression. J Invest Dermatol 1991; 97:624-628.
82. Ho KKL, Halliday GM, Barnetson RSC. Sunscreens protect epidermal Langerhans cells and Thy-1 + but not local contact sensitization from the effects of ultraviolet light. J Invest Dermatol 1992; 98:720-724.
83. Wolf P, Donawho CK, Kripke ML. Effect of sunscreens on UV radiation-induced enhancement of melanoma growth in mice. J Natl Cancer Inst 1994; 86:99-105.
84. Reeve VE, Boehm-Wilcox C, Bosnic M et al. Differential photoimmunoprotection by sunscreens is unrelated to epidermal *cis* urocanic formation in hairless mice. J Invest Dermatol 1994; 103:801-806.
85. Morison WL. The effect of a sunscreen containing para-aminobenzoic acid on the systemic immunologic alterations induced in mice by exposure to UVB radiation. J Invest Dermatol 1984; 83:405-408.
86. Wolf P, Yarosh DB, Kripke ML. Effects of sunscreens and a DNA excision repair enzyme on ultraviolet radiation-induced inflammation, immune suppression, and cyclobutane pyrimidine dimer formation in mice. J Invest Dermatol 1993; 101:523-527.
87. Wolf P, Donawho CK, Kripke ML. Analysis of the protective effect of different sunscreens on ultraviolet radiation-induced local and systemic suppression of contact hypersensitivity and inflammatory responses in mice. J Invest Dermatol 1993; 100:254-259.
88. Morison WL, Pike RA, Kripke ML. Effect of sunlight and its component wavebands on contact hypersensitivity in mice and guinea pigs. Photodermatol 1985; 2:195-204.
89. Roberts LK, Beasley DG. Commercial sunscreen lotions prevent ultraviolet-radiation-induced immune suppression of contact hypersensitivity. J Invest Dermatol 1995; 105:339-344.
90. Roberts LK, Beasley DG, Learn DB et al. Ultraviolet spectral energy differences affect the ability of sunscreen lotions to prevent ultraviolet-radiation-induced immunosuppression. Photochem Photobiol 1996; 63:874-884.
91. Morison WL, Kelley SP. Sunlight suppressing rejection of 280- to 320-nm UV-radiation induced skin tumors in mice. J Natl Cancer Inst 1985; 74:525-527.
92. Walker SL, Young AR. Sunscreens offer the same UVB protection factors for inflammation and immunosuppression in the mouse. J Invest Dermatol: 1997; 108:133-138.
93. Miyagi T, Bhutto AM, Nonaka S. The effects of sunscreens on UVB

erythema and Langerhans cell depression. J Dermatol 1994; 21:645-651.

94. Donawho CK, Wolf P, Kripke ML. Enhanced development of murine melanoma in UV-irradiated skin: UV dose response, waveband dependence, and relation to inflammation. Melanoma Res 1994; 4:93-100.
95. Donawho CK, Kripke ML. Lack of correlation between UV-induced enhancement of melanoma development and local suppression of contact hypersensitivity. Exp Dermatol 1992; 1:20-26.
96. Allen JM, Gossett CJ, Allen SK. Photochemical formation of singlet molecular oxygen in illuminated aqueous solutions of several commercially available sunscreen active ingredients. Chem Res Toxicol 1996; 9:605-609.
97. Learn DB, Beasley DG, Giddens LD et al. Minimum doses of ultraviolet radiation required to induce murine skin edema and immunosuppression are different and depend on the ultraviolet emission spectrum of the source. Photochem Photobiol 1995; 62:1066-1075.
98. Wolf P, Kripke ML. Sunscreens and immunosuppression. J Invest Dermatol 1996; 106:1152-1153.
99. Mommas AM, Van Praag MCG, Bavinck JNB et al. Analysis of the protective effect of topical sunscreens on the UVB-radiation-induced suppression of the mixed-lymphocyte reaction. J Invest Dermatol 1990; 95:313-316.
100. Van Praag MCG, Out-Luyting C, Claas FHJ et al. Effect of topical sunscreens on the UV-radiation-induced suppression of the alloactivating capacity in human skin in vivo. J Invest Dermatol 1991; 97:629-633.
101. Edwards EKJ, Edwards EKS, Frank BL et al. The effect of an ultraprotective sunscreen on Langerhans cell alteration induced by ultraviolet light in human skin. Int J Dermatol 1986; 25:327-329.
102. Elmets CA, Vargas A, Oresajo C. Photoprotective effects of sunscreens in cosmetics on sunburn and Langerhans cell photodamage. Photodermatol Photoimmunol Photomed 1992; 9:113-120.
103. Whitemore SE, Morison WL. Prevention of UVB-induced immunosuppression in humans by a high sun protection factor sunscreen. Arch Dermatol 1995; 131:1128-1133.
104. LeVee G, Oberheiman JL, Cowell K et al. Effect of sunscreen on human immunosuppression by simulated solar radiation. J Invest Dermatol 1995; 104:600 (abstr.).
105. Krien PM, Moyal D. Sunscreens with broad-spectrum absorption decrease the *trans* to *cis* photoisomerization of urocanic acid in the human stratum corneum after multiple UV light exposures. Photochem Photobiol 1994; 60:280-287.
106. Ley RD, Applegate LA. Ultraviolet radiation-induced histopathologic changes in the skin of the marsupial *Monodelphis domestica*. II. Quantitative studies of the photoreactivation of induced hyperplasia and sunburn cell formation. J Invest Dermatol 1985; 85:365-367.

107. De Rijcke S, Heenen M. Decrease of ultraviolet-induced DNA injury in human skin by *p*-aminobenzoic acid esters. Dermatologica 1989; 179:196-199.
108. Lowe NJ, Breedings J. Evaluation of sunscreen protection by measurement of epidermal DNA synthesis. J Invest Dermatol 1980; 174:181-182.
109. Van Prag MC, Roza L, Boom BW et al. Determination of the photoprotective efficacy of a topical sunscreen against UVB-induced DNA damage in human epidermis. J Photochem Photobiol B 1993; 19:129-134.
110. Freeman S, Ley RD, Ley KD. Sunscreen protection against UV-induced pyrimidine dimers in DNA of human skin in situ. Photodermatol 1988; 5:243-247.
111. Hodges NDM, Moss SH, Davies DJG. The sensitizing effect of a sunscreen agent, p-aminobenzoic acid, on near UV-induced damage in a repair deficient strain of *Escherichia coli*. Photochem Photobiol 1977; 26:493-498.
112. Knowland J, McKenzie EA, McHugh PJ et al. Sunlight-induced mutagenicity of a common sunscreen ingredient. FEBS Lett 1993; 324:309-313.
113. Shimoi K, Nakamura Y, Noro T et al. Methyl cinnamate derivatives enhance UV-induced mutagenesis due to the inhibition of DNA excision repair in *Escherichia coli* B/r. Mutat Res 1985; 146:15-22.
114. Sutherland BM. P-aminobenzoic acid sunlamp sensitization of pyrimidine dimer formation and transformation in human cells. Photochem Photobiol 1982; 36:95-97.
115. Sutherland JC, Griffin KP. P-Aminobenzoic acid can sensitize the formation of pyrimidine dimers in DNA: direct chemical evidence. Photochem Photobiol 1984; 40:391-394.
116. Long SD, Little JB. Sunscreen agents induce DNA repair activity in mouse embryo fibroblasts. J Environ Pathol Toxicol Oncol 1984; 5:193-200.
117. Hruza LL, Pentland AP. Mechanisms of UV-induced inflammation. J Invest Dermatol 1993; 100(suppl):35-41.
118. Kligman LH, Akin FJ, Kligman AM. The contributions of UVA and UVB to connective tissue damage in hairless mice. J Invest Dermatol 1985; 84:272-276.
119. Sterenborg HJCM, Van der Leun JC. Tumorigenesis by a long wavelength UV-A source. Photochem Photobiol 1990; 51:325-330.
120. Autier P, Dore JF, Lejeune F et al. Cutaneous malignant melanoma and exposure to sunlamps or sunbeds: An EORTC multicenter case-control study in Belgium, France and Germany. Int J Cancer 1995; 58:809-813.

CHAPTER 8

Sunscreen SPF Values and Immune Protection Levels Are Equivalent When Tests Are Conducted by Appropriate Methods and Procedures

Lee K. Roberts and Douglas B. Learn

Introduction

Exposure to terrestrial solar (sunlight) ultraviolet radiation (UV) is a primary environmental human health hazard. Reports of increased incidence of skin cancer and reduced levels of UV-filtering atmospheric ozone have raised public awareness and concerns about the health risks associated with UV exposure. Public health education programs focused on reducing the risk of solar UV overexposure advocate avoiding mid-day sun exposure, wearing protective clothing and sunglasses, and applying sunscreen to exposed skin.[1] Studies have shown that sunscreens prevent UV-induced sunburn (erythema),[2] photoaging,[3,4] episodes of recurrent herpes labialis (cold sores),[5] DNA damage to skin cells,[6-8] development of precancerous actinic keratoses,[9,10] as well as initiation[11,12] and promotion[13] of skin cancers. There is confusion, however, regarding the efficacy of sunscreens to prevent UV-induced immunosuppression, due to conflicting results reported in various published studies.[14-41] Although it has been suggested that sunscreens provide no protection for the various immunological endpoints affected by UV exposure,[14,15,20-24,28,29,34] there are substantial data that show sunscreens provide protection for the same immunological endpoints,[16-19,23-27,29,34,36,41] in the same experimental models,[16,18,23,24,26-34,36-40] and sometimes within the same reports.[14,15,23,24,29,34]

Sunscreen Photobiology: Molecular, Cellular and Physiological Aspects, edited by Francis P. Gasparro.

Studies that question the ability of sunscreens to prevent immunosuppression gain the most notoriety, but the main body of evidence, i.e., results presented in 24 of the 27 published studies, demonstrates the effective immune protective benefit of sunscreens. Among these data are evidence that sunscreens provide partial to complete protection against the following UV-induced conditions: depletion of dendritic epidermal T-cells[24] and Langerhans cells (LC),[15,24,29,30] depression of epidermal antigen-presenting cell function,[30] suppression of local[26,32-34,36,37] and systemic[16,23,26,31,38,40] contact hypersensitivity (CH) responses, suppression of delayed type hypersensitivity reactions,[27] suppression of nonmelanoma tumor immunity,[16,18,23,39] and activation of tumor-antigen specific suppressor T-cells[14,16,39] in mice; suppression of systemic CH responses in guinea pigs;[17] as well as depletion of Langerhans' cells (LC),[19,25] depression of epidermal antigen-presenting cell function,[41] and suppression of local CH responses[35] in humans. A cursory review of these data does not provide much insight into reasons for the confusion about the capacity of sunscreens to protect the immune system. Thus, we felt a comprehensive analysis of these data and experimental protocols was required to bring some resolution to the question of sunscreen immune protection efficacy.

Careful scrutiny of published results reveals that much of the confusion about sunscreen immune protection efficacy stems from a fundamental misunderstanding of sunscreen formulations and attributes, and the use of nonvalidated, poorly characterized experimental models and testing procedures. The approach and methods employed in most studies significantly deviate from the U.S. Food and Drug Administration's (FDA)[42] prescribed testing procedures used by manufacturers to establish sun protection factors (SPF) for labeling marketed sunscreen products for human use. A number of these critical variables have adversely influenced the results and conclusions published in many of the sunscreen immune protection studies. First, most studies were conducted with UV sources that do not emit a UV power spectra similar to that of sunlight.[14-16,19,20,22-32,34,35,37] Second, UV exposure doses based on nonimmunological endpoints were used to test immunological parameters,[26-32] and exceeded the protective capacity of the sunscreens being tested.[21,23,26-32,34] Third, many studies evaluated simple, noncommercial sunscreen formulations that had low SPF values ≤15[14-18,21,22,26-28,31,32,34,38] and/or consisted of a single active sunscreen ingredient.[14-18,24,26-32,34,38] These simple sunscreen formulations do not have the same UV-absorbing characteristics of most currently marketed sunscreen products and were not appropriately tested on humans to establish their true SPF values.[24-28,30-32,34,38] Fourth, in most studies arbitrary amounts of sunscreen were applied for immune protection testing,[14-19,25,27,28,30-32,38] rather than the measured amount of 2 mg/cm^2 specified for SPF testing.[42] Finally, the basis of the derivation and technical aspects of sunscreen SPF were ignored in designing and performing experiments and in interpreting the results from sunscreen immune protection studies. Collectively, these factors have contributed to inappropriate conclusions drawn from experimental data and inappropriate extrapolation of experimental results to humans. Unlike the experimental conditions employed in most sunscreen immune protection studies,[14-16,19,20,22-32,34,35,37,38] humans use marketed sunscreen products and are only exposed to terrestrial solar UV.

The purpose of this chapter is to review the results of our studies that established the immune protective capacity of marketed sunscreen lotions using experimental protocols[33,36,37,39,40] designed to closely adhere to the procedures outlined for sunscreen SPF testing.[42] Our studies critically evaluated the effects of UV spectra and exposure doses on the immune protective capacity of marketed sunscreen lotions with labeled SPF ranging from 4 to 45 in experimental mouse models of UV-induced LC depletion (Beasley DG et al, submitted for publication), local[33,36,37,40] and systemic[40] suppression of CH responses, and suppression of nonmelanoma tumor-immunity.[39] Dose-response curves were generated in the different immunosuppression models for each UV source that was evaluated. Sunscreen immune protection testing was conducted with defined immunosuppression doses (ISD) derived from each of the UV source dose-response curves. Sunscreen lotions were applied at 2 mg/cm^2 as outlined in the methods for sunscreen SPF testing in humans.[42]

We also outline in this chapter the general approach and procedures for more accurate measurements of sunscreen immune protection factors (IPF), the mathematical equations and methodologies used for calculating IPF, and the relationship between sunscreen IPF determined in experimental animal studies and sunscreen SPF determined in human clinical tests. By controlling critical experimental variables, as well as understanding and employing relevant methodologies based on those used for human sunscreen SPF tests, we demonstrated that the IPF of marketed sunscreen lotions established in mice are consistent with the level of protection predicted by their labeled SPF.

The Basics of Sunscreen Protection

SPF Labeling of Marketed Sunscreen Products

Sunscreens marketed to the public are formulated to reduce the amount of sunburn-causing (erythemogenic) solar UV energy that reaches the protected skin of individuals exposed to sunlight. A sunscreen's labeled SPF thus indicates the level of protection from erythemogenic solar UV it provides. Expressed mathematically, SPF is inversely related to the transmission of effective erythemogenic UV energy through the sunscreen to human skin,[43] as shown here in Equation 8.1:

$$\text{SPF} = \frac{1}{\text{Fraction of Transmitted Erythemogenic UV Energy}}$$

For example, using this equation, if a sunscreen allows 1/15 of the effective erythemogenic UV energy to be transmitted to sunscreen-protected skin its SPF = 15. Figure 8.1 illustrates the relationship, based on this equation, between sunscreen SPF and transmission of erythemogenic UV energy. The curve in Figure 8.1 is plotted as the percent transmission of erythemogenic UV energy versus sunscreen SPF ranging from 1 to 50.

Two caveats must be considered to understand both how SPF values are derived for sunscreens and the intended meaning of SPF for labeling sunscreen products for human use under normal environmental conditions of exposure to UV from sunlight. The first caveat relates to sunscreen testing, in that, the SPF for labeling marketed sunscreen products must be determined by a prescribed method.[42] This method requires: (1) employing a UV source (solar simulator) with a UV spectral power output similar to that of sunlight at mid-day; (2) conducting tests with individuals classified as skin types I, II or III by the Fitzpatrick scale;[44] (3) evenly applying the sunscreen product to the skin at a density of 2 mg/cm^2; (4) applying the sunscreen at least 15 minutes before UV exposure; (5) establishing the minimum erythema dose (MED), defined as the lowest UV dose to produce redness reaching the borders of the exposure site when evaluated 22 to 24 hours after exposure, for each individual in the test by giving a geometric series of UV doses (each separated

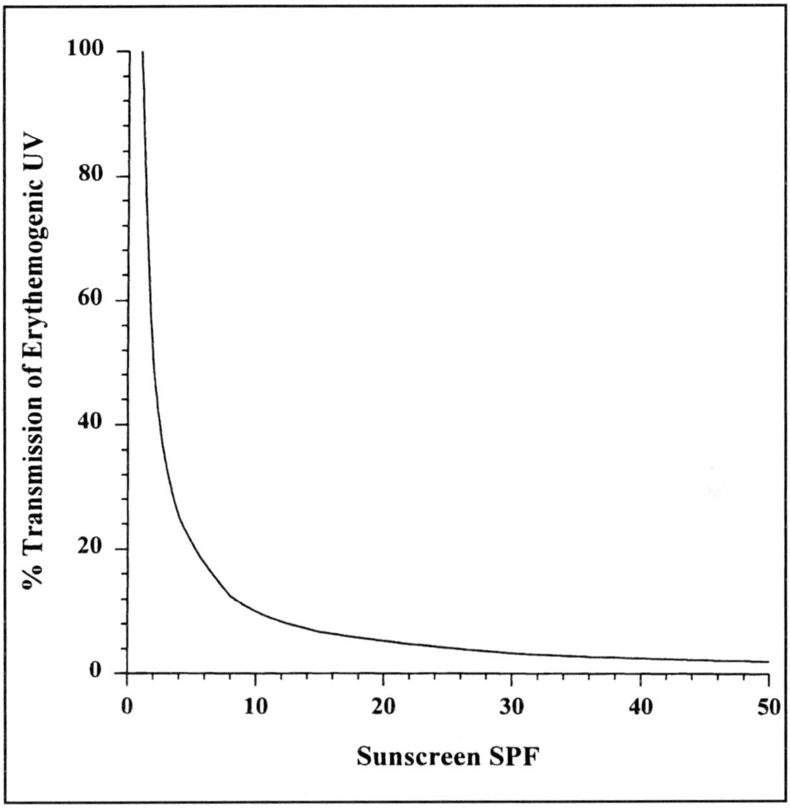

Fig. 8.1. Relationship between sunscreen SPF and transmission of erythemogenic UV energy. The curve was fitted to data generated for SPF values 2, 4, 8, 15, 30 and 45 using Equation 8.1 as described in the text.

by 25% increments) to unprotected skin; and (6) determining the MED on sunscreen-protected skin using a specified geometric series of UV doses, with the UV dose at the middle exposure site based on the sunscreen's estimated SPF, which is used as the multiplier factor for calculating the anticipated MED. Data obtained by this procedure are used to calculate sunscreen SPF according to the formula shown here in Equation 8.2:

$$\text{SPF} = \frac{\text{MED for Sunscreen - Protected Skin}}{\text{MED for Unprotected Skin}}$$

Any deviation from the method outlined above can affect the SPF of a given sunscreen product. For example, a marketed sunscreen product's SPF can be significantly changed if a nonsolar UV source is used, since FDA-approved sunscreen active ingredients used in commercial sunscreen products were selected and formulated to block the transmission of effective erythemogenic solar spectral UV energy. The erythemogenic effectiveness UV spectrum is shifted toward shorter wavelengths if the UV source used for sunscreen testing emits sufficient amounts of nonsolar spectrum UV power.[45,46] Most sunscreens are less efficient at absorbing the highly erythemogenic, but environmentally irrelevant, nonsolar spectrum UV energy at wavelengths below 295 nm. Thus, the resulting SPF of a sunscreen determined with a nonsolar spectrum UV source is effectively lower than the SPF determined with a solar spectrum UV source.

The second caveat relates to the types of protection that sunscreens can provide. The term SPF, by definition, only applies to sunscreen prevention of human erythema by the attenuation of erythemogenic solar UV energy.[42,43] Although sunscreens provide protection against other types of UV damage,[3-13,16-19,23-27,29-41] the levels of nonsunburn protection they provide must be established for each photobiological endpoint. That is, the level of sunscreen protection for a given photobiological endpoint will be relative to, but possibly independent of, the sunscreen's SPF. Therefore, tests to establish nonsunburn protection levels for sunscreens must closely follow the procedures used for determining SPF, including using solar simulator UV exposure doses and biological endpoints appropriate for the particular photobiological response of interest.

We specifically designed our experimental mouse sunscreen immune protection studies to parallel the testing procedures used for human SPF testing of sunscreen formulations. Several Coppertone™ waterproof sunscreen lotions (SPF 4-45), that are marketed in the U.S., were tested for their ability to prevent LC depletion, suppression of local and systemic CH responses, and suppression of tumor immunity. The IPF determined for these sunscreen lotions for each of the immunological endpoints were compared to the levels of protection predicted by their labeled SPF.

In our studies, sunscreen IPF were calculated using the same mathematical relationships used to determine SPF.[36,37,39,40] Like SPF (Equation 8.1), a sunscreen's IPF is inversely related to the amount of immunosuppressive UV energy it allows to be transmitted to protected skin, as shown in Equation 8.3:

$$\text{IPF} = \frac{1}{\text{Fraction of Transmitted Immunosuppressive UV Energy}}$$

In our tests that evaluated prevention of suppression of local[36,37,40] and systemic[40] CH responsiveness, IPF for the different sunscreen lotions were calculated using Equation 8.4:

$$\text{IPF} = \frac{\text{ISD to Sunscreen - Protected Mice}}{\text{ISD to Unprotected Mice}}$$

This is analogous to Equation 8.2 that is used to calculate sunscreen SPF based on human MED given to unprotected and sunscreen-protected skin sites. Similar to SPF calculations based on human MED, the ISD used in Equation 8.4 refers to a specified level of immunosuppression in mice. Equivalent immunosuppression endpoints are used to determine effective ISD for both sunscreen-protected and unprotected mice.

UV Spectral and Dose Relationships Between Mouse Skin Edema and Various Immunosuppression Endpoints

Numerous studies have incorrectly used minimum erythema or skin edema dose (MEdD) to establish sunscreen IPF in mice.[21,23,26-32,34] These studies underestimate the immune protective capacity of sunscreens because the mouse MEdD is significantly higher than minimal ISD.[46] In our studies we defined ISD relative to the immunological endpoint under investigation.[33,36,37,39,40] The rationale we used for selecting minimal ISD used for sunscreen immune protection testing follows; the following sections review each of the experimental immunosuppression models. Dose response comparisons for induction of skin edema and depression of different immunological endpoints in C3H mice exposed to solar simulator UV are presented in Figure 8.2. These data make readily apparent the broad distribution of solar simulator UV doses required to induce suppression of the various immunological endpoints. Furthermore, the solar simulator UV doses needed to induce murine immunosuppression are significantly lower than that for inducing skin edema (Fig. 8.2). Similar differences between immunosuppression and skin edema dose response curves were also observed for nonsolar spectral UV sources (data not shown).

The majority of sunscreen immune protection studies have been conducted with various types of fluorescent sunlamps. Unfiltered UVB sunlamps are the most common UV source used in these studies.[14-16,19,21-23,25-29,31-36,39] Some investigators have used UVB sunlamps or a mixture of UVB and UVA sunlamps filtered with Kodacel-film to reduce the amount of shorter wavelength UV (below 290 nm) in

Fig. 8.2. UV dose-response relationships between immunosuppression and edema in mice. C3H mice were exposed to a range of UV doses from a solar simulator. Dose response comparisons are made for UV-induced depletion of LC, suppression of local and systemic CH, suppression of tumor-immunity and back skin edema. The symbols for each of the photobiological endpoints are given in the figure.

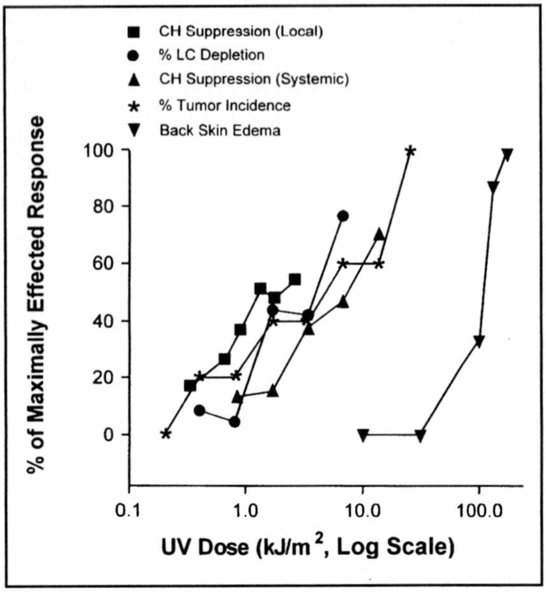

the emission spectrum.[24,33,34,36,39] We have compared the effects of UV spectrum on sunscreen immune protection by conducting tests with unfiltered and Kodacel-filtered UVB sunlamps,[33,36,39] UVA-340 sunlamps (Q-Panel, Cleveland, OH) that emit a near-solar UV power spectrum,[37] and an appropriately filtered 1000 W xenon arc lamp solar simulator.[33,36,37,39,40] In Figure 8.3 the UV power spectra of these UV sources are compared to that of mid-day sunlight. Differences in the relative amounts of nonsolar spectrum UV power emitted by each UV source, i.e., at wavelengths between 250-294 nm, is presented in Table 8.1. Although the amount of nonsolar spectrum UV power emitted by these sources is relatively low compared to their total UV power output, it represents a significant portion of their respective immunosuppression effectiveness spectra (Table 8.1). The influence of nonsolar spectrum UV on induction of immunosuppression and efficacy of sunscreen immune protection will become apparent in the following reviews of each immunological endpoint.

Data in Table 8.2 compares the ISD for each immunological endpoint with the MEdD derived from the dose response curves generated with the different UV sources. The MEdD is significantly higher than both the ISD and the maximum immune suppressive UV dose for all immunological endpoints tested. For mice exposed to solar simulator UV the MEdD is about 3-fold higher than the tumor

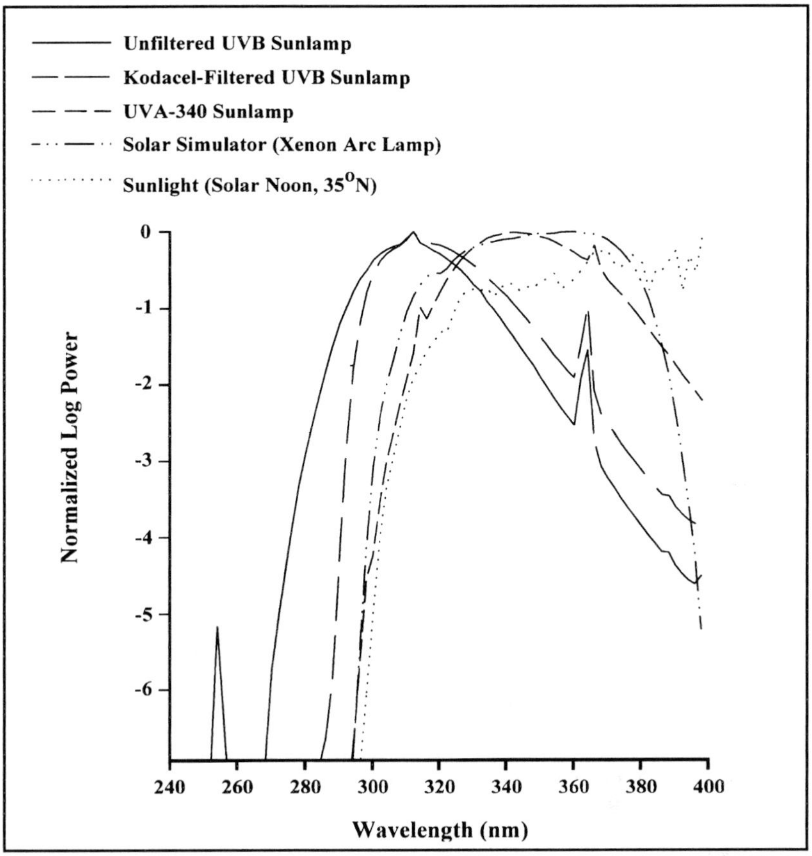

Fig. 8.3. Comparison of UV power spectra. Normalized UV power spectra are given for: a bank of nine Westinghouse FS20 sunlamps (unfiltered UVB sunlamps); the bank of FS20 sunlamps filtered with a single sheet of 0.005 inch thick #K6808 Kodacel cellulose triacetate film (Kodacel-filtered UVB sunlamps); a bank of six Q-Panel UVA-340 sunlamps (UVA-340 sunlamps); a 1000 W ozone-free xenon arc lamp with ellipsoidal reflector, dichroic mirror, 1 mm WG320 filter and 1 mm UG11 filter (solar simulator); and sunlight measured at solar noon, June 18, 1991, at 35° North latitude. Spectra were measured at 2 nm intervals from 250 to 400 nm with an Optronics 742 spectroradiometer using a Teflon dome diffuser. To measure the spectra of the four artificial UV sources the detector was positioned in the center of the irradiation field at the level of the backs of the mice. To derive the relative power scale the actual power (W/cm^2) was converted to a normalized scale by assigning the wavelength with the highest power measurement a value of 1 and adjusting the other wavelength power measurements accordingly.

Table 8.1. Distribution of solar versus non-solar spectrum UV power and effective immunosuppressive UV energy emitted by different UV sources.

UV Source	UV Power*		Effective Immunosuppressive UV Energy+	
	250-294 nm	295-400 nm	250-294 nm	295-400 nm
Unfiltered UVB Sunlamps	7.6%#	92.4%	36.3%	63.6%
Kodacel-filtered UVB Sunlamps	0.5%	99.5%	3.5%	96.5%
UVA-340 Sunlamps	0.1%	99.9%	0.8%	99.2%
Solar Simulator	<0.01%	>99.99%	0.02%	99.98%
Sunlight	<0.001%	>99.999%	~0%	~100%

*Distribution of nonsolar (wavelengths 250-294 nm) and solar (wavelengths 295-400 nm) spectrum UV power for each UV source.
+Distribution of effective immunosuppressive nonsolar and solar spectrum UV energy for each UV source. Immunosuppression effectiveness spectra were derived by multiplying the Elmets et al.[53] multi-hit UV action spectrum for local suppression of CH by the power spectra of each UV source (Fig. 8.4).
#Data are presented as the percentage of the total UV power or effective immunosuppressive UV energy distributed in the nonsolar and solar spectral regions of each UV source.

ISD, 10-fold higher than the ISD for LC depletion and systemic suppression of CH responsiveness, and 50-fold higher than the ISD for local suppression of CH responsiveness. Similar differences between MEdD and the various ISD also hold for unfiltered and Kodacel-filtered UVB sunlamps (Table 8.2). These data help explain why in those studies conducted with nonsolar spectrum UV sources and UV exposures based on MEdD, rather than immunological endpoints, sunscreens appear to provide less protection, than expected, against UV-induced immunosuppression. That is, when UV exposure doses are based on MEdD, the effective ISD greatly exceeds the immune protective capacities of most sunscreens tested, especially those with low SPF values.[21,23,26-34,36,39,40,46]

Investigators that use skin edema or erythema as the biological endpoint to test sunscreen immune protection argue they are

Table 8.2. Comparison of UV immunosuppression and skin edema doses in C3H mice exposed to UV from different sources.

Effective UV Dose*	UV Source+			
	FS	KFS	UVA-340	Solar Simulator
50% Suppression of Local CH[33,36,40]	0.25#	0.9	1.0	1.35
50% Suppression of Systemic CH[40]	ND	ND	ND	6.76
80% LC Depletion[33]	ND	1.8	ND	6.76
100% Tumor Incidence[39]	5.0	7.5	ND	25.3
Back Skin MEdD[46]	1.0	2.8	ND	70.0

*Rationale for selecting ISD is discussed in the reviews of each sunscreen immune protection model. MEdD is defined as the lowest UV dose given to induce back skin edema that is significantly different ($p < 0.05$) from the mean back skin thickness measured on non-UV-irradiated controls.
+UV sources are: unfiltered UVB sunlamps (FS), Kodacel-filtered UVB sunlamps (KFS), UVA-340 sunlamps (UVA-340), and a filtered xenon arc lamp solar simulator (solar simulator).
#UV doses are expressed in kJ/m^2, measured with a IL1700 radiometer and SED 400 detector (International Light, Newburyport, MA). The sensitivity of the detector varies ~12% from its midpoint response at 300 nm over the spectral range 150-400 nm. ND, indicates the test was not performed to determine the specified ISD or MEdD.

correlating sunscreen immune protection levels with SPF.[21,23,26-32,34] However, this argument is invalid for two reasons. First, mouse skin MEdD, and MED,[47] are significantly higher than even the maximum ISD for a full spectrum UV source (Fig. 8.2 and Table 8.2).[46] Second, there is a misconception about the relationship between human and mouse MED, and their use in sunscreen SPF testing. Cole et al[47] showed that MEdD for Skh-1 hairless mice exposed to solar simulator UV are significantly lower than their MED. They,[47] and others,[38] have concluded that mouse MEdD, but not MED, correlates best with human MED. Unlike the MEdD used in sunscreen immune protection studies,[21,23,26-32,34] the Skh-1 mouse skin MEdD reported by Cole et al[47] was defined as the lowest solar simulator UV dose to induce visually perceptible skin swelling in half the irradiated mice. Animals were also restrained during irradiation to assure the solar simulator's UV rays were directed perpendicular to the exposed skin site.[47,48] Studies[21,23,26-32,34] in which mice were unrestrained during irradiation and where caliper measurements were used to determine skin edema results in MEdD that are much higher than those reported by Cole et al.[47,48] Using their specified procedures and definitions of MEdD, Cole et al[48] showed that sunscreen SPF tests conducted in Skh-1 mice correlated well with SPF determined in human clinical tests. However, no other mouse models have been validated against human results for accuracy in determining sunscreen SPF. Since there is no direct relationship between UV-induced skin edema, erythema and immune suppression in mice, past studies using nonimmunological endpoints for sunscreen immune protection testing have resulted in inappropriate conclusions about the relationship between IPF and SPF.

Review of Sunscreen Immune Protection Studies

Sunscreen Protection of Langerhans Cells

Epidermal LC are the primary antigen-presenting cells in the skin.[49] UV-irradiated skin sites are immunologically compromised, in part, because UV causes both depletion of the epidermal LC population,[50] due to their migration from the epidermis to draining lymph nodes,[51] and inactivation of LC antigen-presenting function for inducing effective cell-mediated immune responses.[52] Previous studies have shown sunscreens provide partial to complete protection against LC depletion in the skin of mice[15,24,29,30] and humans[19,25] exposed to nonsolar spectrum UV from sunlamps. We found that marketed sunscreen lotions prevent LC depletion in the skin of mice exposed to a fixed dose of UV from Kodacel-filtered UVB sunlamps (Beasley DG et al, submitted for publication). In that study LC density was reduced about 75% in the skin of unprotected or placebo lotion- (i.e. a lotion without active sunscreen ingredients) treated UV-exposed Skh-1 hairless mice. LC were not depleted in the skin of similarly UV-exposed Skh-1 mice that were protected with an SPF 30 sunscreen. In tests conducted with C3H mice the levels of protection against LC depletion provided by SPF 4, 8, 15, 30 and 45 sunscreen lotions were proportional to their labeled SPF values (Table 8.3). In an earlier study, Roberts and Beasley[33] showed that C3H mice protected with an SPF 15 sunscreen lotion mounted normal CH responses when exposed to a UV dose from Kodacel-filtered UVB sunlamps that caused 84% reduction in LC density and 79% suppression of local CH responsiveness in unprotected mice. Although the SPF 15 sunscreen provided complete immune protection against UV-induced local immunosuppression of CH,[33] it provided only partial protection against LC depletion (Table 8.3). This is consistent with our dose-response data (Fig. 8.2 and Table 8.2); in that, because the

Table 8.3. *Levels of LC protection provided by sunscreens are proportional to their labeled SPF values.*

Treatment*	LC Density+
Non-UV-Irradiated	1320
Unprotected, UV-Irradiated	26 (98%)
SPF 4 Protected, UV-Irradiated	185 (86%)
SPF 8 Protected, UV-Irradiated	343 (74%)
SPF 15 Protected, UV-Irradiated	462 (65%)
SPF 30 Protected, UV-Irradiated	845 (36%)
SPF 45 Protected, UV-Irradiated	1016 (23%)

*Epidermal sheets for evaluation of LC densities were obtained from the skin of non-UV-irradiated control C3H mice, and unprotected or sunscreen-protected C3H mice that were exposed to two consecutive daily doses of UV (1.8 kJ/m^2/day) from Kodacel-filtered UVB sunlamps.[33]
+LC densities were determined by examination of epidermal sheets that were immunoperoxidase stained for Ia-positive cells. Data are presented as the number of LC/mm^2 of epidermis and, in parentheses, percent reduction in LC density compared to the normal LC density in the epidermis of non-UV-irradiated controls.

ISD for local suppression of CH is significantly lower than that for LC depletion, one would predict that at UV doses where sunscreens provide complete protection against immunosuppression of CH they would only provide partial protection against LC depletion. These findings also provide both evidence that UV-induced suppression of CH responses and LC depletion are independent immunological endpoints, and strong support of our argument that to establish accurate sunscreen immune protection values for specific immunological parameters the UV exposure dose must be matched with the immunological endpoint under investigation.[36,37,39,40]

Sunscreen Prevention of UV-Induced Suppression of CH Responses

Some studies have suggested that sunscreens are ineffective at preventing local[15,24,34] and systemic[21,23] suppression of CH responses in mice. However, these studies were conducted with unfiltered[15,21,23] or Kodacel-filtered[24,34] sunlamps that emit high amounts of immunosuppressive UV energy in nonsolar regions of the UV spectrum (Table 8.1);[46] evaluated low SPF (4-8) sunscreen preparations,[21,34] some consisting of a single active sunscreen ingredient;[15,34] and used excessive UV exposure doses based on mouse skin edema[23,34] or erythema[21,23] rather than a defined minimal ISD to test sunscreen immune protection efficacy. Our studies were thus designed to critically evaluate the effects of UV spectra and dose on the capacity of marketed sunscreen lotions to prevent immunosuppression of CH responses.[36,37,40] Minimum ISD, defined as the lowest UV dose required to suppress the mean CH response by 50% in groups of unprotected, UV-irradiated mice compared to that of the nonUV-irradiated controls, was used to conduct sunscreen immune protection tests. The half-maximal CH response, i.e., 50% immunosuppression, was chosen as the endpoint to define ISD because it is consistent with the endpoint used to derive UV action spectra for local[53] and systemic[54] suppression of CH responses. Interestingly, about 50% is the maximum degree of suppression of CH responsiveness observed in mice exposed to solar simulator UV, although CH responses in mice exposed to UV from unfiltered and Kodacel-filtered UVB sunlamps,[36,40] can be suppressed by greater than 80-90%. A differential effect on induction of immunosuppression by these UV sources was also observed in the relative differences among their ISD (Table 8.2). The range of ISD for these UV sources differed by 5-fold, i.e., the ISD are: 0.25 kJ/m^2 for unfiltered UVB sunlamps, 0.9 kJ/m^2 for Kodacel-filtered UVB sunlamps, 1 kJ/m^2 for UVA-340 sunlamps, and 1.35 kJ/m^2 for a filtered xenon arc lamp solar simulator. These data[36,37,40,46] clearly demonstrate the enhanced immunosuppressive effectiveness of nonsolar spectrum UV, at wavelengths between 250-294 nm, from these UV sources (ranging from 0.1 to 7.6% of the total UV power; Table 8.1) compared to that of solar spectrum UV delivered by a solar simulator (representing 0.01% of the total UV power; Table 8.1).

An extensive series of experiments were conducted, using a protocol patterned after the method for sunscreen SPF testing,[42] to the determine IPF of four marketed sunscreen lotions labeled SPF 4, 8, 15 and 30 for prevention of local[36] suppression of CH. Immune protection data were used to calculate each sunscreen's IPF using the previously described Equation 8.4. The IPF of each sunscreen (Table 8.4) exceeded the level of protection predicted by its labeled SPF in tests conducted with the solar simulator. Conversely, sunscreen IPFs were significantly less than that predicted by their labeled SPF in tests conducted with unfiltered and Kodacel-filtered UVB sunlamps (Table 8.4). Interestingly, in tests conducted with UVA-340 sunlamps,[37] which emit only a slight amount of nonsolar

Table 8.4. Effect of UV power spectrum on measured sunscreen IPF for preventing local immunosuppression of CH responses.

	Immune Protection Factor*			
Sunscreen Lotion⁺	**FS**	**KFS**	**UVA-340**	**Solar Simulator**
SPF 4	1	4	8	15
SPF 8	2	8	15	15
SPF 15	4	8	15	30
SPF 30	4	8	30	60

*IPF were calculated using Equation 8.4 and ISD data obtained by previously described methods.[36,37] The UV sources are described in the footnotes to Table 8.2.
⁺Coppertone™ waterproof sunscreen lotions identified by SPF were evaluated in these studies. Sunscreen immune protection testing procedures have been previously described.[36,37]

spectrum UV power, sunscreen IPF were equivalent to the level of protection predicted by their labeled SPF, but were lower than those determined in tests conducted with the solar simulator (Table 8.4).[36] Thus, in addition to immunosuppressive effectiveness, the relatively small amounts of nonsolar spectrum UV, at wavelengths between 250 to 294 nm, emitted by these different sunlamps have a significant influence on the calculated IPF of a sunscreen lotion.

Comparison of immunosuppression effectiveness spectra provides graphic proof of the extent to which nonsolar spectrum UV energy from fluorescent sunlamps contributes to the induction of suppression of CH responses in mice (Fig. 8.4). Immunosuppression effectiveness spectra presented in Figure 8.4 were derived by multiplying the Elmets et al[53] multi-hit UV action spectrum for induction of local suppression of CH responsiveness in C3H mice with the UV power spectra of sunlight and each of the UV sources. It is clear from these immunosuppression effectiveness spectra that wavelengths of UV below 295 nm emitted by unfiltered and Kodacel-filtered UVB sunlamps, and UVA-340 sunlamps, but not a solar simulator or sunlight, make a significant contribution toward the induction of CH immunosuppression. Integration of these effectiveness spectra shows that 36.3%, 3.5% and 0.8% of the total effective immunosuppressive UV energy emitted by unfiltered UVB sunlamps,

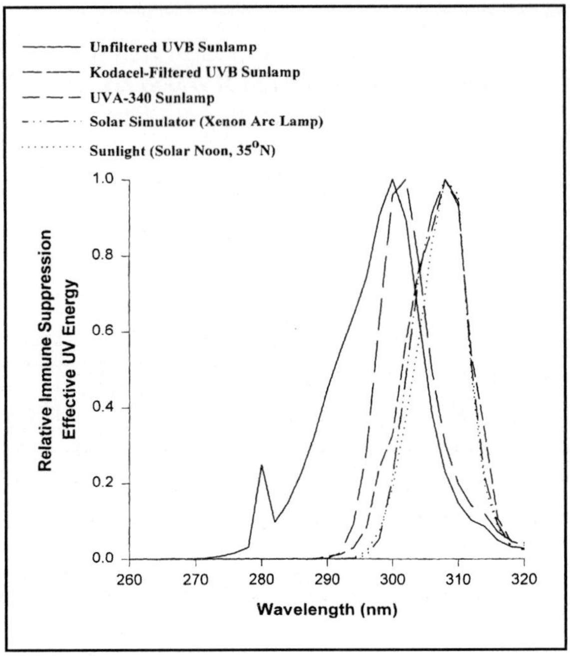

Fig. 8.4. Comparison of immunosuppression effectiveness spectra. Immunosuppression effectiveness spectra were derived by multiplying the UV power spectrum of each UV source (Fig. 8.3) by the Elmets et al[53] multi-hit UV action spectrum for local suppression of CH in C3H mice. The effectiveness spectra were normalized as described in the legend for Figure 8.3.[36] Portions of this figure are reprinted with permission from Roberts LK, Beasley DG, Learn DB et al. Photochem Photobiol 1996; 63:874-884. © 1996 American Society for Photobiology.

Kodacel-filtered UVB sunlamps, and UVA-340 sunlamps, respectively, is contributed by the nonsolar spectral regions of their UV power spectra (Table 8.1). Conversely, only 0.02% of the total immunosuppressive UV energy produced by a filtered xenon arc lamp solar simulator lies below 295 nm (Table 8.1). These data dramatically illustrate the strong immunosuppressive effect of relatively small amounts of nonsolar UV power in the spectra of fluorescent sunlamps, that is not present in the UV power spectra of either sunlight or an appropriate solar simulator.

The strong immunosuppressive effects of nonsolar spectrum UV energy also influence the immune protection efficacy of sunscreens. Figure 8.5 shows the spectra of the transmission of effective immunosuppressive UV power through an SPF 30 sunscreen lotion for unfiltered and Kodacel-filtered UVB sunlamps, UVA-340 sunlamps, a solar simulator, and sunlight. These spectra reveal the widely differing proportions of effective immunosuppressive UV power from the nonsolar UVC (wavelengths 200-290 nm) and short, nonsolar UVB (wavelengths 290-294 nm) spectral regions of each UV source that is transmitted through a sunscreen. The nonsolar spectrum UV power from fluorescent sunlamps that is transmitted through sunscreens make a significant contribution toward the induction of immunosuppression. For example, 20.2%, 2.5% and 0.8% of the total immunosuppressive UV energy from unfiltered UVB sunlamps, Kodacel-filtered UVB sunlamps, and UVA-340 sunlamps, respectively, that is transmitted through an SPF 30 sunscreen is from the nonsolar regions of their UV power spectra. However, less than 0.01% of the immunosuppressive UV energy from a solar simulator that is transmitted through an SPF 30 sunscreen lies below 295 nm. Differences in the relative amounts of nonsolar spectrum immunosuppressive UV energy and the overall magnitude of immunosuppressive UV energy from each UV source that reaches sunscreen-protected skin (Fig. 8.5) further illustrates why there is considerable variability in sunscreen IPF that are measured with these various UV sources and how testing with nonsolar spectrum UV sources can underestimate the immune protection capacity of sunscreens.

Sunscreens Provide Tumor-Immune Protection

UV is a known carcinogen for the induction of skin cancer in humans and experimental animals. In addition to the direct carcinogenic and tumor-promoting effects of UV, there is evidence that UV-induced immunosuppression plays an active role in the emergence and progression of mouse skin tumors.[55] Unlike chemically-induced skin tumors, the majority of skin cancers induced in mice by UV are immunologically rejected when transplanted to secondary syngeneic recipients.[56] These "UV-regressor tumors" will grow, however, when transplanted into immunologically compromised hosts or syngeneic mice exposed to subcarcinogenic UV doses.[56] The apparent immunological defect in the tumor immune surveillance system of UV-exposed mice is mediated by a population of tumor-antigen specific suppressor T-lymphocytes (Ts-cells) that are activated in UV-exposed mice before the appearance of neoplastic skin cells.[57] UV-induced Ts-cells are elicited by cross-reactive common

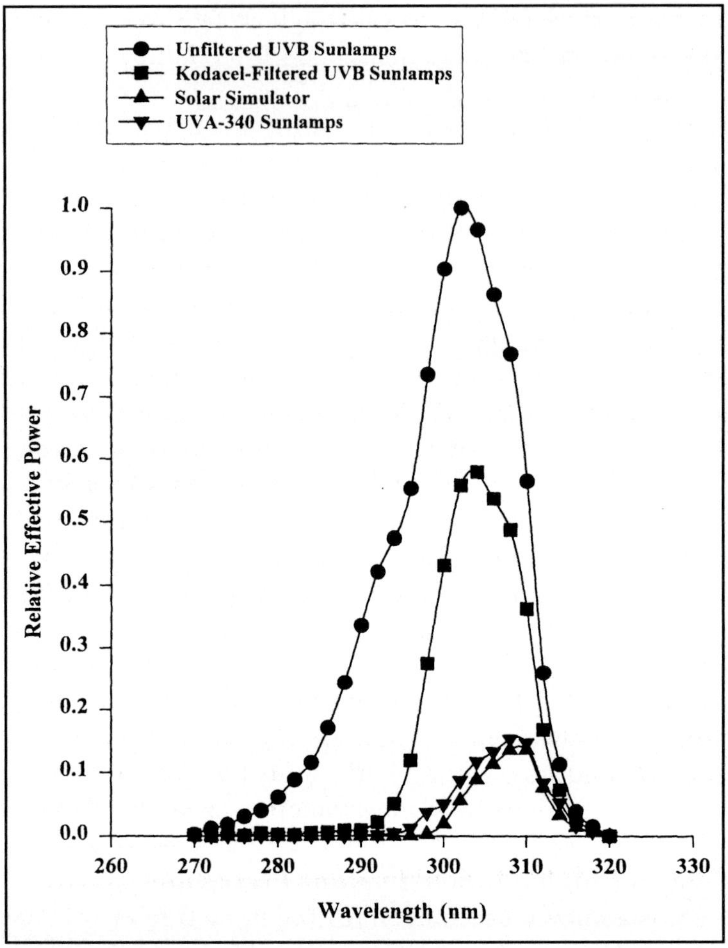

Fig. 8.5. Comparison of sunscreen transmission spectra of different UV source immunosuppression effectiveness spectra. Immunosuppressive UV effective power transmission spectra through an SPF 30 sunscreen lotion are compared for the different UV sources. The spectra are presented as the relative effective UV power transmitted through the sunscreen over the UV spectral range 270-320 nm, which covers the range of the immunosuppression effectiveness spectra shown in Figure 8.4. Portions of this figure are reprinted with permission from Roberts LK, Beasley DG, Learn DB et al. Photochem Photobiol 1996; 63:874-884. ©1996 American Society for Photobiology.

tumor-associated-antigens (TAA). These TAA are expressed in UV-irradiated skin and by virtually all skin cancers.[58,59] Once activated, this population of TAA-specific Ts-cells remains functional for an extended time even after UV exposure is discontinued.[60] The role of these Ts-cells in photocarcinogenesis is to both permit the emergence of neoplastic cells[61] and promote the growth of skin cancers by preventing tumor rejection responses mediated by cytotoxic T-cells.[57-59] Thus, the experimental mouse model for evaluating UV-induced tumor-immune suppression involves analysis of both tumor incidence and growth rate of UV regressor tumors that are transplanted into either UV-exposed secondary syngeneic hosts[56] or nonUV-irradiated recipients of an adoptive transfer of TAA-specific Ts-cells.[57]

Sunscreens have been shown to be effective at preventing UV-induced carcinogenesis by blocking DNA damage in human skin,[6-8] preventing initiation[11,12] and promotion[13] of skin cancer in mice, inhibiting the development and persistence of precancerous actinic keratoses in humans,[9,10] and reducing skin cancer risk in xeroderma pigmentosum patients.[62] However, there are conflicting reports regarding the ability of sunscreens to prevent UV-induced tumor immune suppression in mice.[14,16,18,23,28,39] As with the other immunological endpoints discussed in this review, most sunscreen tumor immune protection studies were conducted with unfiltered[14,16,23,28,39] or Kodacel-filtered[39] UVB sunlamps; used relatively high UV exposure doses based on endpoints of skin edema,[23,28] erythema,[23] or photocarcinogenesis;[14,16] and tested low SPF sunscreen preparations,[16,18,28] some containing a single active sunscreen ingredient.[14,16,18,28] Morison reported that a marketed SPF 8 sunscreen lotion, containing a single active sunscreen ingredient, only provided partial protection against suppression of tumor immunity in mice exposed to UV from unfiltered UVB sunlamps.[16] Interestingly, the same sunscreen lotion provided complete tumor immune protection for mice exposed to UV from sunlight.[18] As with other immunological models, these data[16,18] suggest that the effects of UV dose and nonsolar spectrum UV on reducing the apparent efficacy of sunscreens to provide tumor immune protection were likely.

Similar to our studies using the models of UV-induced CH suppression, we took a systematic approach to evaluate the effect of UV spectrum and dose on sunscreen tumor immune protection.[39] Tumor immune suppression UV dose-response curves were derived with unfiltered and Kodacel-filtered UVB sunlamps, and a solar simulator (Fig. 8.6). Tumor immune suppression dose (TISD), defined as the lowest dose tested to induce susceptibility to the outgrowth of UV regressor tumor transplants in 100% of unprotected UV-exposed mice, were derived from the dose-response curves for each UV source. TISD for unfiltered (5 kJ/m^2) and Kodacel-filtered (7.5 kJ/m^2) UVB sunlamps differed by only 1.5-fold, but were 5- and 3-fold lower, respectively, than the solar simulator TISD (25.3 kJ/m^2).[39]

Sunscreen tumor immune protection tests were conducted with fixed TISD of each UV source. The tumor immune protective capacity of SPF 8, 15, 30 and 45 marketed sunscreen lotions were tested. Both transplant tumor incidence and growth rates were significantly reduced in sunscreen-protected versus unprotected mice exposed to UV from a solar simulator.[39] Sunscreen tumor IPF, calculated with the tumor incidence data, were found to match the levels of protection predicted by each of the sunscreen's labeled SPF in tests conducted with a solar simulator (Table 8.5). The SPF 30 and 45 sunscreens were also found to block the activation of TAA-specific Ts-cells in mice exposed to TISD of solar simulator UV.[39] In contrast, sunscreen tumor-IPF were significantly lower in tests conducted with Kodacel-filtered UVB sunlamps (Table 8.5). None of the sunscreens provided measurable tumor immune protection for mice exposed to TISD of UV from unfiltered UVB sunlamps (Table 8.5). Thus, nonsolar spectral UV energy from unfiltered and Kodacel-filtered UVB sunlamps influence both the UV dose required to induce tumor immune suppression (Fig. 8.6) and the tumor immune protective capacity of sunscreens (Table 8.5). And, as was observed in the models of immunosuppression of CH,[33,36,37,40] sunscreen tumor IPF matched the levels of protection predicted by their labeled SPF when testing procedures were based on TISD and employed a solar simulator UV source[39] that emitted a UV power spectrum similar to that of sunlight.[46]

Sunscreen Prevention of Immunosuppression in Humans

The preponderance of experimental data generated in animal models demonstrate the ability of sunscreens to prevent UV-induced immunosuppression.[15-18,23,24,26,27,29-34,36-40] However, the most important question is whether marketed sunscreen products protect humans against immunosuppression induced by exposure to UV from sunlight; and whether a marketed sunscreen's labeled SPF predicts

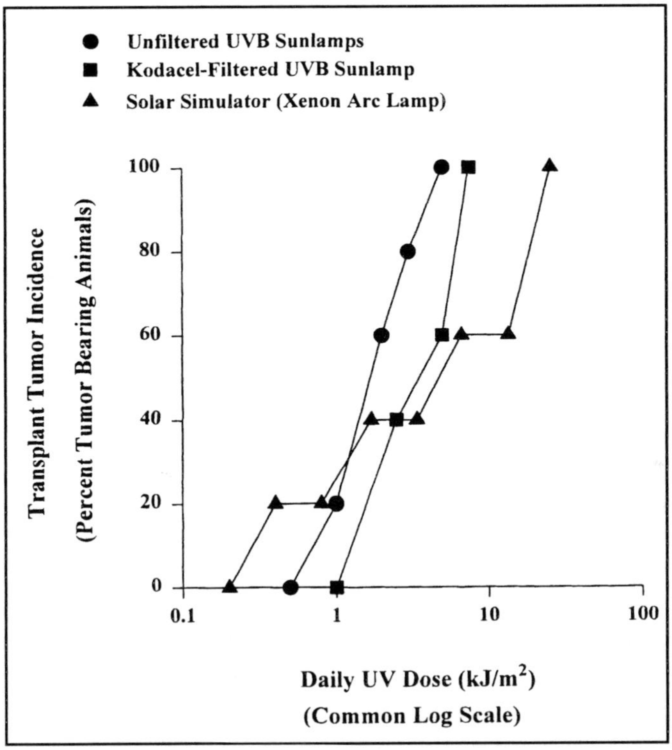

Fig. 8.6. Tumor immune suppression dose-response curves for three UV sources. Groups of C3H mice were exposed to a range of UV doses delivered from unfiltered and Kodacel-filtered UVB sunlamps and a solar simulator (UV sources are described in Fig. 8.3). Data are presented as the percentage of tumor-bearing mice per total number of mice in each UV exposure group (transplant tumor incidence) for each UV dose point measured three weeks after implantation of a UV-regressor tumor. Portions of this figure are reprinted with permission from Roberts LK, Beasley DG. Int J Cancer 1997; 71:94-102. © 1997 Wiley-Liss, Inc.

its immune protective capacity. Recent studies suggest that this is indeed the case. Whitmore and Morison[35] showed that a marketed SPF 29 sunscreen lotion provided complete immune protection against UV-induced suppression of CH responsiveness in skin type II and III individuals that were exposed to 3 MED of UV from unfiltered UVB sunlamps. LeVee et al[63] reported that CH responses, while suppressed in unprotected individuals exposed to 4 MED of solar simulator UV, were normal in similarly UV-exposed individuals that were protected with a marketed SPF 4 sunscreen lotion. Despite that

Table 8.5. Effect of UV power spectrum on measured sunscreen IPF for preventing tumor immune suppression.

	Immune Protection Factor*		
Sunscreen Lotion+	**FS**	**KFS**	**Solar Simulator**
SPF 8	1	1.1	8
SPF 15	1	1.9	15
SPF 30	1	2.2	30
SPF 45	NT	7.3	45

*IPFs were calculated according to previously described methods.[39] The UV sources are described in the footnotes to Table 8.2. NT indicates that the test was not done.
+Coppertone™ waterproof sunscreen lotions identified by SPF were evaluated in these studies. Sunscreen tumor immune protection testing procedures have been previously described.[39]
Reprinted with permission from Roberts LK, Beasley DG. Int J Cancer 1997; 71:94-102. © 1997 Wiley-Liss, Inc.

sensitizations were made through sunburned skin of the sunscreen protected individuals that received an effective 1 MED, their CH responses were higher than those observed in unprotected individuals that were exposed to UV doses less than 1 MED.[63] Further clinical investigation is needed to confirm these findings; however, these results clearly demonstrate that sunscreens do provide immune protection for humans[35,63] and that SPF is predictive of the level of immune protection provided by a sunscreen when individuals are exposed to solar spectrum UV.[63]

Conclusions

Although some reports question the ability of sunscreens to protect the immune system from the effects of UV,[14,15,20-24,28,34] the preponderance of experimental animal,[15-18,23,24,26,27,29-34,36-40] human clinical,[19,25,35,63] and in vitro[41] data demonstrate that sunscreens prevent UV-induced immunosuppression. Our studies[33,36,37,39,40] have identified the key testing parameters required for accurate assessment of sunscreen immune protection in mice. The correct UV source and appropriate immunosuppressive UV dose are the most critical factors for conducting accurate sunscreen immune protection tests. We showed nonsolar UV spectral power, at wavelengths between 250-294 nm, emitted by filtered and unfiltered fluorescent sunlamps is highly immunosuppressive.[36,46] Most sunscreen immune protection studies have been conducted with unfiltered UVB sunlamps[14-16,19,21-23,25-29,31-36,39] or Kodacel-filtered sunlamps.[24,33,34,36,39] These UV sources produce significant amounts of effective immunosuppressive nonsolar spectrum UV energy at wavelengths below 295 nm (Table 8.1). Even UVA-340 sunlamps that emit very little nonsolar spectrum UV (Fig. 8.3 and Table 8.1) produce ISD for local suppression of CH and sunscreen IPF that are lower than those of an appropriately filtered xenon arc lamp solar simulator (Tables 8.2 and 8.4).[37]

Like others,[15,16,19,23-27,29-36] we showed sunscreens can provide some protection against immunosuppression induced by UV from unfiltered and Kodacel-filtered UVB sunlamps.[33,36] However, the degree of sunscreen immune protection observed in tests conducted with these UV sources is limited. The immune protective capacity of certain sunscreen formulations, especially those with low SPF, are easily exceeded when mice are exposed to relatively high doses of filtered and unfiltered sunlamp UV. Conversely, sunscreens provide adequate levels of immune protection in tests conducted with solar spectral UV from either a solar simulator[33,36,37,39,40,63] or sunlight.[17,18] Morison and colleagues showed that an SPF 8 sunscreen lotion provided high levels of immune protection for animals exposed to UV from sunlight,[17,18] but only partial protection in tests conducted with unfiltered UVB sunlamps.[16] We demonstrated that the IPF of marketed sunscreens for preventing suppression of local[36] and systemic[40] CH responses, and tumor immunity[39] in mice equal or exceed the level of protection predicted by their labeled SPF. The fact that sunscreen IPF can be calculated indicates doses of solar simulator UV can be given that exceed sunscreen immune protective capacities. This is readily apparent from the Fisher et al[21] study which reported that an SPF 6 sunscreen, although capable of blocking erythema, did not prevent suppression of systemic CH responses in mice exposed to erythemogenic doses (72 kJ/m^2/day) of solar simulator UV. The erythemogenic dose of UV used in that study[21] was >10-fold higher than the ISD (6.76 kJ/m^2/day; Table 8.4) obtained from our dose response curve.[40] It's not surprising that the SPF 6 sunscreen failed to provide immune protection in the Fisher et al[21] study, since we have shown that mice protected with an SPF 8 sunscreen mount normal systemic CH responses after exposure to 8 ISD (54.1 kJ/m^2), but not 15 ISD (101.4 kJ/m^2), of solar simulator UV.[40] Thus, like Fisher et al,[21] our data[36,40] demonstrate that immunosuppression is inducible in sunscreen-protected mice that are exposed to doses of solar simulator UV which exceed the immune protection capacity of the sunscreen. These data further demonstrate the need to use solar spectrum UV doses based on a specified immunological endpoint rather than erythema (or edema) when evaluating sunscreen immune protection in mice, due to the large discrepancy between ISD and MEdD (Fig. 8.2 and Table 8.1). Walker and Young[38] recently reported that sunscreen protection factors for prevention of inflammation (edema) and immunosuppression were the same. A narrow band UV source, 92% of the total UV power was at 311 ± 4 nm, and an experimental protocol where the mouse MEdD and ISD were equivalent, were used in this study.[38]

The difference between ISD and MED might be less for humans. LeVee et al[63] found that CH responses were only partially depressed in sunscreen-protected versus unprotected individuals exposed to an effective MED of solar simulator UV. That is, despite the fact that the sunscreen-protected individuals were sensitized through sunburned skin they were able to mount an effective CH response. Thus in humans, sunscreen SPF appears to be an adequate predictor of protection against both sunburn and UV-induced immunosuppression of CH.

Our data clearly demonstrate the need to conduct sunscreen testing experiments with an appropriate immunosuppressive UV dose delivered with a UV source that produces an environmentally relevant UV energy spectrum, i.e., sunlight.[33,36,37,39,40,46] These findings strongly indicate a need to critically re-examine the results and conclusions that have been drawn about sunscreen immune protection in past studies conducted with nonsolar spectrum UV sources. Our data[33,36,37,39,40,46] also indicate the need to validate experimental models and methods, based on accepted clinical SPF testing procedures,[42] before extrapolating sunscreen immune protection results to humans. Neither accurate measurements of sunscreen efficacy nor appropriate extrapolation of experimental results to humans are possible without adhering to these requirements .

Acknowledgments

We thank our collaborators Donathan G. Beasley, John Beard, Susan Edmonds, Lori D. Giddens, Stephen J. Moloney, Mary A. Montgomery and Joseph W. Stanfield for their contributions to the work presented in this chapter.

References

1. Marks R. Primary prevention of skin cancer. Br Med J 1994; 309:285-286.
2. Pathak MA. Sunscreens: topical and systemic approaches for protection of human skin against harmful effects of solar radiation. J Am Acad Dermatol 1982; 7:285-312.
3. Kligman LH. Protective effect of sunscreens against photoaging. Drug Cosmet Ind 1985; 137:30-34.
4. Kligman LH, Akin FJ, Kligman AM. Sunscreens promote repair of ultraviolet radiation-induced dermal damage. J Invest Dermatol 1983; 81:98-102.
5. Rooney JF, Bryson Y, Mannix ML et al. Prevention of ultraviolet-light-induced herpes labialis by sunscreen. Lancet 1991; 338:1419-1422.
6. Freeman SE, Ley RD, Ley KD. Sunscreen protection against UV-induced pyrimidine dimers in DNA of human skin in situ. Photodermatol 1988; 5:243-247.
7. De Rijcke S, Heenen M. Decrease of ultraviolet-induced DNA injury in human skin by p-aminobenzoic acid esters. Dermatologica 1989; 179:196-199.
8. van Praag MCG, Roza L, Boom BW et al. Determination of the photoprotective efficacy of a topical sunscreen against UVB-induced DNA damage in human epidermis. J Photochem Photobiol B: Biol 1993; 19:129-134.
9. Thompson SC, Jolley D, Marks R. Reduction of solar keratoses by regular sunscreen use. N Engl J Med 1993; 329:1147-1151.
10. Naylor MF, Boyd A, Smith DW et al. High sun protection factor sunscreens in the suppression of actinic neoplasia. Arch Dermatol 1995; 31:170-175.
11. Kligman LH, Akin FJ, Kligman AM. Sunscreens prevent ultraviolet photocarcinogenesis. J Am Acad Dermatol 1980; 3:30-35.
12. Forbes PD, Davies RE, Sambuco CP et al. Inhibition of ultraviolet radiation-induced skin tumors in hairless mice by topical application of the sunscreen 2-ethyl hexyl-4'-methoxycinnamate. J Toxicol Cut Ocular Toxicol 1989; 8:209-226.
13. Reeve VE, Bosnic M, Boehm-Wilcox C. Effect of ultraviolet (UV) radiation and UVB-absorbing sunscreen ingredients on 7,12-dimethylbenz(a)anthracene-initiated skin tumorigenesis in hairless mice. Photodermatol Photoimmunol Photomed 1990; 7:222-227.
14. Gurish MF, Roberts LK, Krueger GG et al. The effect of various sunscreen agents on skin damage and the induction of tumor susceptibility in mice subjected to ultraviolet irradiation. J Invest Dermatol 1981; 76:246-251.
15. Lynch DH, Gurish MF, Daynes RA. Relationship between epidermal Langerhans cell density ATPase activity and the induction of contact hypersensitivity. J Immunol 1981; 126:1892-1897.
16. Morison WL. The effect of a sunscreen containing Para-aminobenzoic acid on the systemic immunologic alterations induced in mice by exposure to UVB radiation. J Invest Dermatol 1984; 83:405-408.

17. Morison WL, Pike RA, Kripke ML. Effect of sunlight and its component wavebands on contact hypersensitivity in mice and guinea pigs. Photodermatol 1985; 2:195-204.
18. Morison WL, Kelley SP. Sunlight suppressing rejection of 280- to 320-nm UV-radiation-induced skin tumors in mice. J Natl Cancer Inst 1985; 74:525-527.
19. Edwards EK Jr, Edwards EK Sr, Frank BL et al. The effect of an ultraprotective sunscreen on Langerhans cell alteration induced by ultraviolet light in human skin. Intl J Dermatol 1986; 25:327-329.
20. Hersey P, MacDonald M, Burns C et al. Analysis of the effect of a sunscreen agent on the suppression of natural killer cell activity induced in human subjects by radiation from solarium lamps. J Invest Dermatol 1987; 88:271-276.
21. Fisher MS, Menter JM, Willis I. Ultraviolet radiation-induced suppression of contact hypersensitivity in relation to padimate O and oxybenzone. J Invest Dermatol 1989; 92:337-341.
22. van Praag MCG, Out-Luyting C, Claas FHJ et al. Effect of topical sunscreens on the UV-radiation-induced suppression of the alloactivating capacity in human skin in vivo. J Invest Dermatol 1991; 97:629-633.
23. Reeve VE, Bosnic M, Boehm-Wilcox C et al. Differential protection by two sunscreens from UV radiation-induced immunosuppression. J Invest Dermatol 1991; 97:624-628.
24. Ho KKL, Halliday GM, Barnetson RSC. Sunscreens protect epidermal Langerhans cells and Thy-1$^+$ cells but not local contact sensitization from the effects of ultraviolet light. J Invest Dermatol 1992; 98:720-724.
25. Elmets CA, Vargas A, Oresajo C. Photoprotective effects of sunscreens in cosmetics on sunburn and Langerhans cell photodamage. Photodermatol Photoimmunol Photomed 1992; 9:113-120.
26. Wolf P, Donawho CK, Kripke ML. Analysis of the protective effect of different sunscreens on ultraviolet radiation-induced local and systemic suppression of contact hypersensitivity and inflammatory responses in mice. J Invest Dermatol 1993; 100:254-259.
27. Wolf P, Yarosh DB, Kripke ML. Effects of sunscreens and a DNA excision repair enzyme on ultraviolet radiation-induced inflammation, immune suppression, and cyclobutane pyrimidine dimer formation in mice. J Invest Dermatol 1993; 101:523-527.
28. Wolf P, Donawho CK, Kripke ML. Effect of sunscreens on UV radiation-induced enhancement of melanoma growth in mice. J Natl Cancer Inst 1994; 86:99-105.
29. Miyagi T, Bhutto AM, Nonaka S. The effect of sunscreen on UVB erythema and Langerhans cell depression. J Dermatol 1994; 21:645-651.
30. Walker SL, Morris J, Chu AC et al. Relationship between the ability of sunscreens containing 2-ethylhexyl-4'-methoxycinnamate to protect against UVR-induced inflammation, depletion of epidermal

Langerhans (Ia$^+$) cells and suppression of alloactivating capacity of murine skin in vivo. J Photochem Photobiol B:Biol 1994; 22:29-36.
31. Reeve VE, Boehm-Wilcox C, Bosnic M et al. Differential photoimmunoprotection by sunscreen ingredients is unrelated to epidermal *cis* urocanic acid formation in hairless mice. J Invest Dermatol 1994; 103:801-806.
32. Wolf P, Cox P, Yarosh DR et al. Sunscreens and T4N5 liposomes differ in their ability to protect against ultraviolet-induced sunburn cell formation, alterations in dendritic epidermal cells, and local suppression of contact hypersensitivity. J Invest Dermatol 1995; 104:287-292.
33. Roberts LK, Beasley DG. Commercial sunscreen lotions prevent ultraviolet radiation induced immune suppression of contact hypersensitivity. J Invest Dermatol 1995; 105:339-344.
34. Bestak R, Barnetson RSC, Nearn MR et al. Sunscreen protection of contact hypersensitivity responses from chronic solar-simulated ultraviolet irradiation correlates with the absorption spectrum of the sunscreen. J Invest Dermatol 1995; 105:345-351.
35. Whitmore SE, Morison WL. Prevention of UVB-induced immunosuppression in humans by a high sun protection factor sunscreen. Arch Dermatol 1995; 131:1128-1133.
36. Roberts LK, Beasley DG, Learn DB et al. Ultraviolet spectral energy differences affect the ability of sunscreen lotions to prevent ultraviolet radiation induced immunosuppression. Photochem Photobiol 1996; 63:874-884.
37. Beasley DG, Beard J, Stanfield JW et al. Evaluation of an economical sunlamp that emits a near solar UV power spectrum for conducting photoimmunological and sunscreen immune protection studies. Photochem Photobiol 1996; 64:303-309.
38. Walker SL, Young AR. Sunscreens offer the same UVB protection factors for inflammation and immunosuppression in the mouse. J Invest Dermatol 1997; 108:133-138.
39. Roberts LK, Beasley DG. Sunscreen lotions prevent ultraviolet radiation-induced suppression of antitumor immune responses. Int J Cancer 1997; 71:94-102.
40. Roberts LK, Beasley DG. Sunscreens prevent local and systemic immunosuppression of contact hypersensitivity in mice exposed to solar simulated ultraviolet radiation. J Photochem Photobiol B:Biol 1997; in press.
41. Degwert J, Pape W, Hoppe U. In vitro analysis of immunoprotective effects of topical sunscreens. Toxic in Vitro 1994; 8:747-749.
42. U.S. Food and Drug Administration, Tentative Final Monograph for sunscreen drug products for over-the-counter human use. Fed Reg 1993; 53(90):28194-28302.
43. Sayre RM, Marlowe E, Agin PP et al. Performance of six sunscreen formulations on human skin. Arch Dermatol 1979; 115:46-49.
44. Fitzpatrick TB. The validity and practicality of sun-reactive skin types I through VI. Arch Dermatol 1988; 124:869-871.

45. Learn DB, Beard J, Moloney SJ. The ultraviolet C energy emitted from FS lamps contributes significantly to the induction of human erythema and murine ear edema. Photodermatol Photoimmunol Photomed 1993; 9:147-153.
46. Learn DB, Beasley DG, Giddens LD et al. Minimum doses of ultraviolet radiation required to induce murine skin edema and immunosuppression are different and depend on the ultraviolet emission spectrum of the source. Photochem Photobiol 1995; 62:1066-1075.
47. Cole CA, Davies RE, Forbes PD et al. Comparison of action spectra for acute cutaneous responses to ultraviolet radiation: man and albino hairless mouse. Photochem Photobiol 1983; 37:623-631.
48. Cole CA, Forbes PD, Ludwigsen K. Sunscreen testing using the mouse ear model. Photodermatol 1989; 6:131-136.
49. Cruz PD, Bergstresser PR. Antigen processing and presentation by epidermal Langerhans cells. Immunodermatol 1990; 8:633-647.
50. Toews GB, Bergstresser PR, Tigelaar RE et al. Epidermal Langerhans cell density determines whether contact hypersensitivity or unresponsiveness follows skin painting with DNFB. J Immunol 1980; 134:445-453.
51. Moodycliffe AM, Kimber I, Norval M. Role of tumour necrosis factor-α in ultraviolet B light-induced dendritic cell migration and suppression of contact hypersensitivity. Immunol 1994; 81:79-84.
52. Tang A, Udey MC. Inhibition of epidermal Langerhans cell function by low dose ultraviolet B radiation. J Immunol 1991; 146:3347-3355.
53. Elmets CA, LeVine MJ, Bickers DR. Action spectrum studies for induction of immunologic unresponsiveness to dinitrofluorobenzene following in vivo low dose ultraviolet radiation. Photochem Photobiol 1985; 42:391-397.
54. Noonan FP, DeFabo EC. Ultraviolet-B dose-response curves for local and systemic immunosuppression are identical. Photochem Photobiol 1990; 52:801-810.
55. Roberts LK, Lynch DH, Samlowski WE et al. Ultraviolet radiation and modulation of the immune response. In: Norris DA, ed. Immune Mechanisms in Cutaneous Disease. New York: Marcell Dekker Inc., 1989:167-215.
56. Daynes RA, Spellman CW, Woodward JG et al. Studies into the transplantation biology of ultraviolet light-induced tumors. Transplantation 1977; 23:343-348.
57. Roberts LK. Characterization of a cloned ultraviolet radiation (UV)-induced suppressor T cell line that is capable of inhibiting anti-UV tumor-immune responses. J Immunol 1986; 136:1908-1916.
58. Roberts LK, Lynch DH, Daynes RA. Evidence for two functionally distinct cross-reactive tumor antigens associated with ultraviolet light and chemically-induced tumors. Transplantation 1982; 33:352-360.
59. Hong SR, Roberts LK. Cross-reactive tumor antigens in the skin of mice exposed to subcarcinogenic doses of ultraviolet radiation. J Invest Dermatol 1987; 88:154-160.

60. Spellman CW, Daynes RA. Properties of ultraviolet light-induced suppressor lymphocytes within a syngeneic tumor system. Cell Immunol 1978; 36:383-387.
61. Fisher MS, Kripke ML. Suppressor T lymphocytes control the development of primary skin cancers in ultraviolet-irradiated mice. Science 1982; 216:1133-1134.
62. Kondoh M, Ueda M, Nakagawa K et al. Siblings with xeroderma pigmentosum complementation group A with different skin cancer development: Importance of sun protection at an early age. J Am Acad Dermatol 1994; 31:993-996.
63. LeVee GJ, Oberheiman L, Cowell K et al. Effect of sunscreen on human immunosuppression by simulated solar radiation. J Invest Dermatol (Abstr) 1995; 104:600.

CHAPTER 9

The Relationship Between Sunscreen Protection from Erythema, Photoimmunosuppression and Photocarcinogenesis in the Hairless Mouse

Vivienne E. Reeve

Introduction

Skin cancer incidence rates continue to rise worldwide in spite of massive education campaigns about the risks of sun exposure, and the ready availability of sunscreens.[1-4] There has been a doubling in the incidence of both nonmelanoma skin cancers, and of malignant melanoma in the State of Queensland, Australia, in the decade to 1995—equal to the rate of new skin cancers reported for the previous decade to 1985. The lag in the achievement of the anticipated downturn in cutaneous tumors in the population may reflect the characteristically prolonged latency period, perhaps 30-40 years, of these tumors, if sunscreens were to act primarily by preventing new tumor initiations. However, continuing solar UV exposure also has a tumor-promoting function, and it is reasonable to expect that sunscreens should protect from the outgrowth or promotion of existing latent tumor cells, and that this form of photoprotection should be manifested more rapidly in the population as a measurable reduction in skin cancer rates. It has therefore become critical to assess the ways in which sunscreen protection from UV radiation-induced skin cancer has been measured in experimental animals, in order to confirm that sunscreens do indeed

Sunscreen Photobiology: Molecular, Cellular and Physiological Aspects, edited by Francis P. Gasparro.

afford protection, not only from sunburn, but also from photocarcinogenesis.

While the mutagenic events that initiate photocarcinogenesis are most effectively caused by the UVB waveband (290-320 nm) of solar radiation, stimulation of the numerous tumor-promotional factors identified to date appears to result from both UVB and UVA (320-400 nm) radiation. The promotion phase of photocarcinogenesis is particularly powerfully regulated by immunological mechanisms in both man[5-9] and mouse,[10,11] and UV radiation is an effective inhibitor of the normal immunological defenses relevant to tumor recognition and rejection. The primary photoreceptor for the immune impairment appears to be epidermal *trans*-urocanic acid,[12,13] which becomes photoisomerized to *cis*-urocanic acid in irradiated skin; the *cis* isomer has immunosuppressive properties similar to those of UVB radiation, which can be observed with exogenous *cis*-urocanic acid applied topically or parenterally to mice. On the other hand, DNA damage,[14,15] histamine,[16] and inflammatory prostaglandins[17] also appear to be relevant mediators.

It has been demonstrated in mice that the action spectrum for photoimmunosuppression, assayed by the delayed or contact hypersensitivity reaction, is maximal at 260-280 nm, indicating that the shortest wavelength UVB radiation component of sunlight, 290-320 nm, will be responsible.[18] Since the erythema-protective UVB-absorbers in topical sunscreens remove precisely this waveband, having peak UV absorbance around 310 nm, it seems likely that sunscreens will protect from photoimmunosuppression also. However, the immunological consequences of exposure to longer UV wavelengths, as occurs when skin is irradiated for prolonged periods through a UVB-absorbing sunscreen, are unclear, and there is evidence both for[19] and against[20] a contributory immunosuppressive role mediated by UVA radiation. Certainly there is currently much controversy surrounding the vexing question of whether, and how efficiently, sunscreens provide immunoprotection. Differing methods of assay in mouse models may account for the various results, but different UV-absorbers may also react differently. We and others have found a differential protection from UV radiation-induced immunosuppression in mice exposed through two different UVB absorbers.[21-25]

In these studies in the Skh:HR hairless mouse, we have compared the protection afforded by two chemical UVB-absorbers,

2-ethylhexyl-p-methoxycinnamate (2-EHMC) and octyl-N-dimethyl-p-aminobenzoate (o-PABA), from UV radiation-induced erythema/edema (measured by an increase in the skinfold thickness at 24 or 48 h postirradiation) and from UV radiation-induced suppression of contact hypersensitivity (CHS). The early experiments used unfiltered UVB radiation, commercial sunscreens with unmatched vehicle composition, and low doses of UVB radiation which would be only minimally erythemal in unprotected mice. However, it has now been demonstrated that the more closely the test radiation simulates the solar UV spectrum, the more effectively sunscreens can be seen to prevent photoimmunosuppression,[26] and the more relevant the assay probably is to human sunlight exposure. In addition, sunscreens are much more likely to be used by humans when prolonged sunlight exposures are anticipated rather than sub- or minimally erythemal exposures. Therefore in successive studies, we have created experimental sunscreens formulated with a common vehicle (an innocuous oil-in-water cosmetic base lotion), used solar-simulated UV radiation (SSUV), and increased the daily exposure doses to the erythema-protecting capacity of the sunscreen. We believe these experimental conditions better approximate the conditions under which Australians experience sunlight exposure through a topical sunscreen.

Commercial Sunscreens and UVB Radiation

A small collection of commercially marketed sunscreening cosmetics with sun protection factors (SPFs) 10 to 15+, based on either 2-EHMC or o-PABA together with the UVA-absorber benzophenone-3, has been examined for photoprotection in the hairless mouse. One product contained both 2-EHMC and o-PABA with benzophenone, and two products contained in addition, the physical UV-blocker, titanium dioxide, one with an alternative UVA-absorber (4-hydroxy-4-methoxy-benzophenone) (Table 9.1). These sunscreens were applied at 0.2 ml per mouse and spread lightly to cover all the dorsal skin areas, and allowed to be absorbed for 30 minutes. Mice were exposed on the dorsum to 1 minimal erythemal dose (MED) daily of unfiltered UVB radiation (1.84 kJ/m^2 UVB, 1.13 kJ/m^2 UVA) from a fluorescent tube source[21] on 3 consecutive days, an irradiation regime which we have found consistently results in marked suppression of CHS in unprotected Skh:HR mice without cumulatively causing more than a mild erythema reaction. Maximum mid-dorsal

Table 9.1. Active ingredients of commercial sunscreens.

Designation	SPF	UVB absorber	UVA absorber
E1	SPF 10	2-EHMC 6.5%	benzophenone-3 2%
E2	SPF 10	2-EHMC 8%	benzophenone-3 2%
E3	SPF 15	2-EHMC	benzophenone-3
E4	SPF 15	2-EHMC 7.5%	benzophenone-3 4.5%
P1	SPF 15	o-PABA 7.0%	benzophenone-3 3%
P2	SPF 15	o-PABA 6.5%	benzophenone-3 3%
P3	SPF 15+	o-PABA	benzophenone-3
P4	SPF 15+	o-PABA 8%	benzophenone-3 3%
EP1	SPF 15	o-PABA 8% + 2-EHMC 4%	benzophenone-3 4%
EP2	SPF 15+	o-PABA 8% + 2-EHMC 7% + TiO_2 2%	benzophenone-3 6%
EP3	SPF 15+	o-PABA + 2-EHMC + TiO_2	4-hydroxy-4-methoxy-benzophenone

skinfold thickness was recorded at 48 h after the first irradiation, and the systemic CHS reaction was induced by topical application of either 3% oxazolone in ethanol, or 0.3% 2,4-dinitrofluorobenzene in acetone at 7 days, and elicited by challenge on the pinnae 7 days later.[20]

Erythema/Edema

The results showed that all the sunscreens prevented erythema/edema compared with a nonsunscreen base lotion, as expected following the minimally erythemal UVB exposures, although two o-PABA sunscreens, P3 and P4, appeared to cause an edematous reaction in the dark, a source of some concern (Table 9.2).

Contact Hypersensitivity (CHS)

Some sunscreens, eg., E2, P1, P2 (Table 9.3) also appeared to suppress CHS in the dark. As this could not be assigned consistently to either UVB-absorber, it must have been a property of the remaining ingredients in the product for which we were unable to control. In fact, the hairless mouse may be a sensitive monitor for topical irritancy of cosmetic ingredients.

Table 9.2. Erythema/edema measured by dorsal skinfold thickness before and after exposure to 3 x 1 MED of unfiltered UVB radiation through commercial sunscreens.

	Skinfold thickness (x 0.01 mm) ± S.D.	
	Unirradiated	**Irradiated (3 x 1 MED)**
No lotion	82.8 ± 8.6	142.0 ± 11.4
Base lotion	83.8 ± 2.0	143.5 ± 14.0
Commercial sunscreens		
E1	89.9 ± 5.3	89.2 ± 5.3
E2	84.1 ± 7.1	89.9 ± 9.7
E3	79.3 ± 6.6	82.3 ± 6.9
E4	89.4 ± 5.1	85.9 ± 5.6
P1	74.1 ± 6.3	82.3 ± 6.3
P2	87.9 ± 2.88	2.8 ± 2.5
P3	98.0 ± 3.6	97.8 ± 6.1
P4	106.2 ± 5.8	103.9 ± 10.9
EP1	87.4 ± 5.6	95.3 ± 8.4
EP2	97.0 ± 9.7	93.0 ± 5.6
EP3	85.0 ± 6.9	87.1 ± 9.8

Under UVB-irradiation conditions (3 x 1 MED) which caused approximately 50% suppression of CHS in unprotected mice (Table 9.3), the products segregated clearly into the 2-EHMC-based sunscreens through which mice maintained normal or even slightly potentiated (E2) CHS responses, and the o-PABA-based sunscreens which failed to protect (57-89% immunosuppression). When o-PABA and 2-EHMC were present together (EP1), substantial (44%) immunosuppression was still observed, suggesting that o-PABA + UVB immunosuppression was established by a mechanism distinct from the pathways protected by 2-EHMC. When titanium dioxide was included (EP2, EP3) there was almost total protection (13%, 9% immunosuppression), suggesting that the microfine particulate titanium dioxide, which reflects and absorbs across a broad UV spectrum, reduced the immunosuppressive hazard by superior reduction of the relevant incident waveband.

Table 9.3. The systemic CHS reaction after topical application of commercial sunscreens, with and without exposure to 3 X 1 MED unfiltered UVB radiation.

Treatment	Average ear swelling x 0.01 mm ± S.E.				% UVB-induced I/S
	Non-irradiated	% of "No lotion"	Irradiated	% of Non-UV Control	
No lotion	24.6 ± 2.3	100%	12.0 ± 1.1	49%	51
Base	23.4 ± 1.4	95	12.7 ± 1.7	54	46
E1	21.1 ± 2.3	86	22.7 ± 1.8	108	-8
E2	13.2 ± 2.8	54	19.1 ± 2.5	145	-45
E3	17.6 ± 2.7	72	18.1 ± 3.9	103	-3
E4	14.7 ± 1.5	60	15.9 ± 1.7	108	-8
P1	13.5 ± 1.8	55	5.8 ± 1.8	43	57
P2	13.5 ± 2.3	55	1.5 ± 1.2	11	89
P3	18.3 ± 2.5	74	9.7 ± 2.0	53	67
P4	19.6 ± 2.3	80	4.3 ± 3.0	22	78
EP1	22.9 ± 2.0	93	12.7 ± 1.4	56	44
EP2	15.2 ± 3.3	62	13.2 ± 1.6	87	13
EP3	14.5 ± 2.0	59	13.2 ± 1.7	91	9

Tumor Susceptibility

A second assessment of the T cell immune status involved comparing the establishment of susceptibility to transplantable UV radiation-induced tumor cells in mice irradiated through sunscreens—either E4 (immune-protective for CHS) or P2 (non immune-protective for CHS).[21] An intradermal inoculum of 10^5 cultured cells resulted in a proliferating tumor at the site in 64-71% of UVB-irradiated mice (Fig. 9.1). If the mice received UVB through E4, the proliferating tumor incidence was reduced to 35%, identical with mice which had received topical E4 but were not irradiated. In contrast, 73% of mice irradiated through P2 supported tumor proliferation, similar to mice irradiated with UVB without this sunscreen (64% tumor incidence). This study therefore confirmed the photoimmunoprotective properties of E4, and the failure of P2 to protect, using an alternate measure of T cell immunity than the CHS reaction.

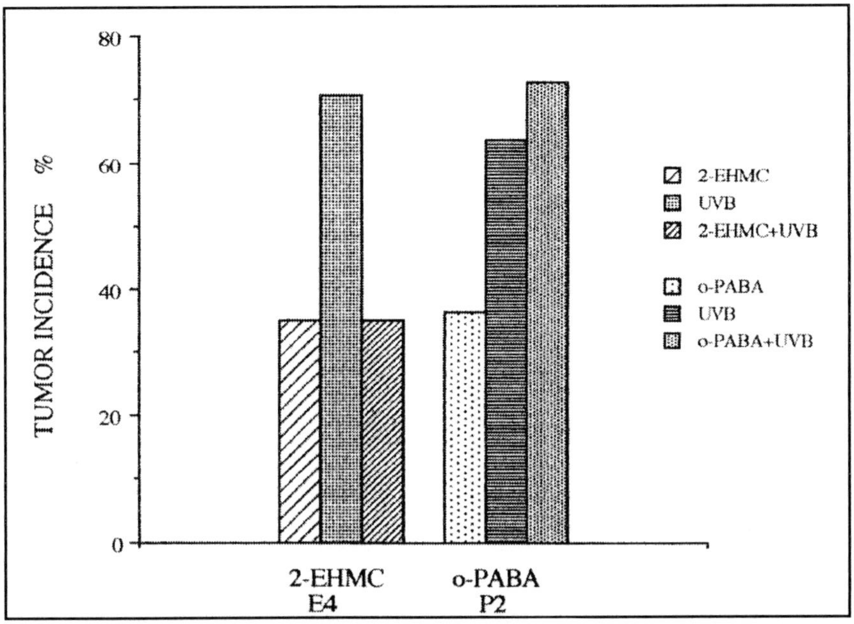

Fig. 9.1. Induction of susceptibility to transplanted tumor cell growth following topical sunscreen application (E4 or P2), with and without exposure to 3 X 1 MED unfiltered UVB radiation.

Urocanic Acid Photoisomerization

Since the UV absorption of *trans*-urocanic acid is maximal at 277 nm, it was of interest to examine the degree of *cis*-urocanic acid formation in sunscreen-irradiated epidermis. We extracted and quantitated the urocanic acid isomers from the epidermis of mice irradiated with 1 MED through E4 (immune-protective) or P2 (non immune-protective), but could not detect any significant isomerization or differences (Table 9.4) between the two UVB-absorbers after irradiation with 3 X 1 MED, although there was an approximately 6-fold increase in *cis* isomer in unprotected mouse skin.

Summary

Thus these studies in commercial products indicated that o-PABA-based sunscreens were inferior to 2-EHMC-based sunscreens in immunoprotection from low dosage UVB radiation (mildly erythemal in unprotected mice), whether the assay was the systemic CHS reaction or the susceptibility to transplanted tumor cell growth. The difference could not be assigned to different degrees of

Table 9.4. Epidermal urocanic acid isomers before and after exposure to 3 X 1 MED of unfiltered UVB radiation through two commercial sunscreens E4 and P2, compared with non-sunscreen-protected mice.

Treatment	μg UCA/cm² ±S.D.			% *cis* UCA
	trans UCA	*cis* UCA	Total	
Nil	7.2 ± 1.0	0.4 ± 0.2	7.7	6
UVB	7.4 ± 2.5	2.5 ± 0.5	9.9	25
E4	6.7 ± 0.9	0.4 ± 0.2	7.2	6
E4 + UVB	5.2 ± 0.7	0.4 ± 0.1	5.7	8
P2	6.1 ± 0.9	0.4 ± 0.1	6.5	7
P2 + UVB	4.9 ± 0.6	0.4 ± 0.1	4.4	10

protection from erythema or from the photoisomerization of urocanic acid. These observations were consistent when experimental sunscreen lotions, formulated with 2-EHMC or o-PABA at various concentrations in a common base lotion, were compared,[22] including exposures to larger UVB radiation doses which approached, or slightly exceeded, the erythema-protective capacity of the sunscreens. It was apparent, in comparison with 2-EHMC, that o-PABA in the same base lotion was significantly less protective from erythema than predicted by its in vitro UVB absorption potential, suggesting that o-PABA became localized differently in the cutaneous strata after topical application, perhaps thus shielding targets for erythema or immunosuppression less effectively than 2-EHMC. Alternatively, there may be relevant differences in the metabolic degradation of the UVB-absorbers, or perhaps chemical reactivity with cellular components.[27,30] Photodegradation products of UV-absorbers have been recently described, the biological effects of which are unknown.[31-36] These might also result in spectral changes in the transmitted UV radiation. Nonetheless, erythema-protectively matched lotions displayed clearly that 2-EHMC readily prevented UVB suppression of CHS, and o-PABA failed. No photoisomerization of urocanic acid was detected with either sunscreen, under conditions where UVB-irradiation through the base lotion produced 23% conversion to the *cis* isomer.

Experimental SPFf 6 Sunscreens and Simulated Solar UV (SSUV) Radiation

A second series of experiments aiming more accurately to reproduce sunlight exposure conditions in sunscreen-wearing humans compared the effects of irradiation with 6 MED of SSUV through 2-EHMC and o-PABA sunscreens formulated in a common vehicle to provide SPF 6 photoprotection in the Skh:HR mouse. The selection of SPF 6 was arbitrary, but restricted by the prolonged irradiation times which a higher SPF would have necessitated. The responses of the mice were compared with mice irradiated with 1 MED of SSUV (2.00 kJ/m^2 UVB, 24.55 kJ/m^2 UVA) through the base lotion alone. Each exposure treatment therefore should have resulted in an effective radiation dose of 1 MED, although sunscreened mice would have achieved their minimal erythema with a UV spectrum relatively deficient in UVB but enriched with UVA.

It has been important also to assess the effect of sunscreen usage defined by these criteria not only on acute (3 X) SSUV exposures but also on chronic irradiation and photocarcinogenesis induction. When sunscreens have been tested for anti-photocarcinogenic efficacy in mice, generally high SPF products have been combined with low daily doses of UV radiation, equating to 1 MED in the unprotected mouse. Under these conditions, the sunscreens invariably demonstrate efficient protection from tumor induction.[37-42] Therefore, in addition to the acute treatments measuring erythema and immunosuppression, treatments were continued 5 days/week for 15 weeks, and the tumor outgrowth monitored.[43]

Contact Hypersensitivity

Irradiation of mice with 3 X 1 MED of SSUV through base lotion resulted in 42% suppression of the CHS response (Fig. 9.2). Mice irradiated with 3 X 6 MED through SPF 6 o-PABA were equally immunosuppressed (42% suppression of CHS), as predicted if the protection from erythema matched the protection from immunosuppression. However, mice irradiated with 3 X 6 MED through SPF 6 2-EHMC were significantly protected, CHS being suppressed by only 12%. These results are in agreement with our previous experiments with unfiltered UVB radiation,[21,22] and underline the consistently superior protection from immunosuppression by 2-EHMC.

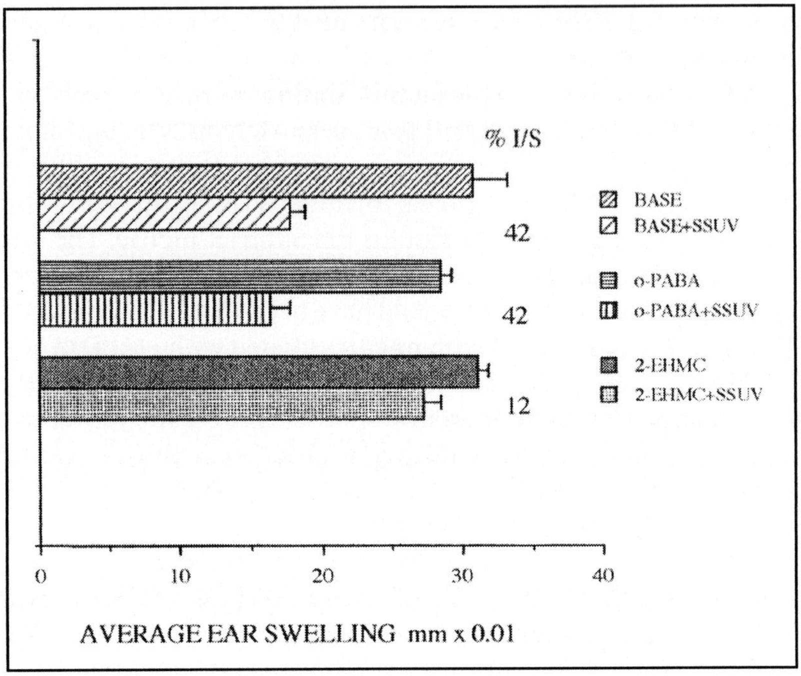

Fig. 9.2. Systemic contact hypersensitivity to oxazolone in mice irradiated with 3 X 1 MED of SSUV through base lotion, or with 3 X 6 MED through SPF 6 sunscreen lotions, compared with nonirradiated mice.

Urocanic Acid Photoisomerization

We found that approximately 22-29% photoisomerization occurred in unprotected epidermis, with no significant difference between 5 X or 20 X SSUV (Table 9.5), suggesting that the photostationary state for the isomerization may have been achieved for hairless mouse skin. Neither UVB-absorber had any significant effect on this isomerization, an interesting observation which differed from previous experiments in which unfiltered UVB irradiation through sunscreens had not resulted in measurable *cis*-urocanic acid formation in the epidermis. The dose of radiation does not seem to have been important, since photoisomerization in response to erythemal unfiltered UVB (3 X 5 MED) radiation had been blocked as efficiently as minimally erythemal UVB (3 X 1 MED).[22]

However, SSUV differs by its large UVA component, to which must be assigned the photoisomerizing potential from which the UVB-absorbers were unable to protect in this study. This is consistent with reports of the unexpected capacity of UVA radiation to

Table 9.5. Urocanic acid extracted from epidermis of normal skin before and following 5 X or 20 X 1 MED SSUV through base lotion, or 5 X or 20 X 6 MED SSUV through sunscreens.

Treatment	[UCA] µg/cm^2 ±S.E.	% cis UCA
No UV	3.9 ± 0.1	4
5 X SSUV exposures:		
Base + SSUV	4.3 ± 0.3	22
o-PABA+SSUV	3.9 ± 0.4	25
2-EHMC+SSUV	4.5 ± 0.5	25
20 X SSUV exposures:		
Base + SSUV	4.9 ± 0.2	29
o-PABA + SSUV	4.6 ± 0.2	22
2-EHMC	5.4 ± 0.5	18

cause photoisomerization in vivo,[20,44,45] although the absorption spectrum of urocanic acid has its peak around 277 nm. The significance of UVA-induced *cis*-urocanic acid is an important question which needs clarification.

It appears then that we have observed no correlation between epidermal urocanic acid photoisomerization and the CHS response in mice irradiated through the sunscreens. The lack of photoisomerization following sunscreened exposure to unfiltered UVB radiation suggests that mediators other than *cis*-urocanic acid may predominate when suppression of CHS has ensued (e.g. through o-PABA sunscreen), mediators which may also be active in the presence of *cis*-urocanic acid arising in response to sunscreened SSUV exposure. However, it is also possible that 2-EHMC in particular might inactivate the epidermal *cis*-urocanic acid in some way, effectively reducing its immunosuppressive capacity.

Cis-Urocanic Acid and UVB-Absorbers

The possibility that 2-EHMC, but not o-PABA, might interfere with *cis*-urocanic acid generated in the skin was examined by topical treatment of mice with sunscreens followed by exogenous *cis*-urocanic acid-containing lotion. It was revealed that whereas topical *cis*-urocanic acid suppressed systemic CHS by 28% in control mice (base lotion only), and by 21% in o-PABA-treated mice, the suppression of CHS in 2-EHMC-treated mice was totally abrogated (Fig. 9.3). If SSUV-generated *cis*-urocanic acid is the immunosuppressive mediator, this finding provides a possible mechanism for enhanced immunoprotection by 2-EHMC. In o-PABA-irradiated mice, the *cis*-urocanic acid activated immunosuppression unhampered. However, the differential in sunscreen protection from UVB-induced immunosuppression, when *cis*-urocanic acid did not appear abundantly in the epidermis, must have resulted from other suppressive pathways. It now remains to identify the relevant reactions between 2-EHMC and *cis*-urocanic acid, whether immediate interactive chemistry is occurring, or whether 2-EHMC acts on downstream reactions normally set in motion by *cis*-urocanic acid. This inactivation of *cis*-urocanic acid by 2-EHMC but not by o-PABA may provide an explanation for the disparate effects of these two UVB-absorbers.

Photocarcinogenesis Induction by Chronic Irradiation Through SPF 6 Sunscreens

Chronic irradiation of groups of mice through the SPF 6 sunscreens, 5 times/week for 15 weeks, produced tumorigenesis as described in a preliminary publication.[43] The final average tumor multiplicities were 48 (base lotion + SSUV), 37 (2-EHMC + SSUV) and 3 (o-PABA + SSUV) tumors per mouse. The 2-EHMC response was approximately as anticipated, and equivalent to the base lotion response, however, the almost total inhibition of tumorigenesis by o-PABA was unexpected. It is difficult to reconcile this superior protection from tumorigenesis by o-PABA with its inferior protection from photoimmunosuppression, unless photoimmunosuppression has no influence on photocarcinogenesis. However, numerous other experimental models support such a relationship, and when photoimmunosuppression in mice has been modified by the anti-inflammatory drug indomethacin,[17] by the histamine H2 receptor antagonist cimetidine,[16] by urocanic acid,[46] by antioxidant polyphenols[47] or by dietary fat alterations,[48,49] photocarcinogenesis has been altered in parallel.[46,49-53] Irradiation through these sunscreens appears to be the exception, and suggests that unidentified sunscreen-dependent chemistry may be responsible, particularly as we had already observed that brief treatment with 2-EHMC alone was able to interfere with the immunosuppressive action of *cis*-urocanic acid.

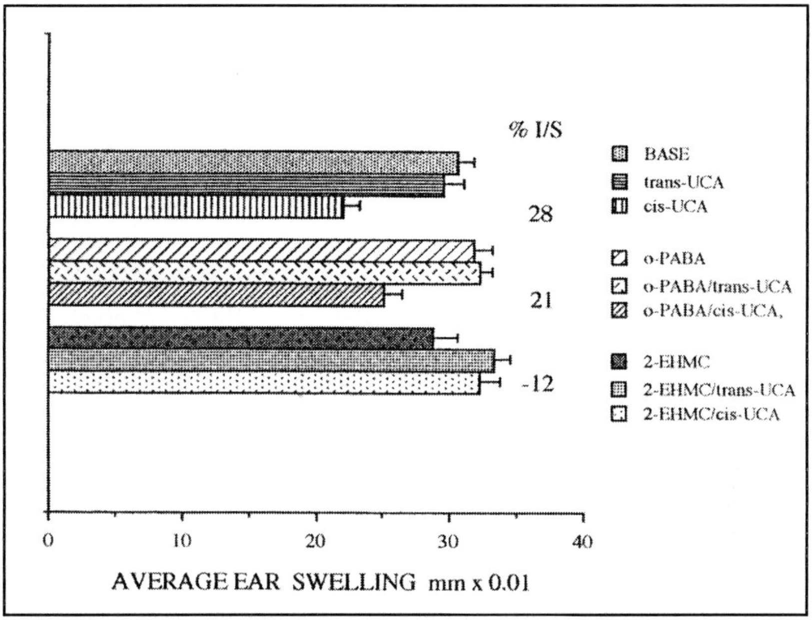

Fig. 9.3. Systemic contact hypersensitivity to 2,4-dinitrofluorobenzene in mice treated topically on 3 consecutive days with base lotion or sunscreen lotions, followed 60 min later with 200 µg *trans*-urocanic acid lotion or 200 µg photoisomerized urocanic acid lotion (containing 51% *trans*-, 49% *cis*-UCA).

Chronic Sunscreen Application and SSUV or Cis-Urocanic Acid

In an attempt to reconcile the opposing effects of o-PABA and 2-EHMC on photoimmunosuppression and photocarcinogenesis, we explored the possibility that repeated sunscreen applications might cause relevant cutaneous changes not seen after acute treatment. The base lotion, SPF 6 o-PABA or 2-EHMC were therefore applied to the mouse dorsum daily X 3, or X 20, followed finally by 3 X 1 MED SSUV (base) or 3 X 6 MED (sunscreens), or treatment with *cis*-urocanic acid lotion, and the effects on the CHS reaction were measured.

Chronic base lotion application had no effect on the suppression of CHS by SSUV exposure (42-48% suppression; Fig. 9.4). Chronic o-PABA application, however, which permitted 22% suppression of CHS after only 3 applications, now appeared to be immunoprotective, and after 20 applications prior to irradiation, the o-PABA mice were immunosuppressed by only 3%. On the other hand, in contrast to the acute application of 2-EHMC, which, as previously found, permitted only 7% suppression of CHS after SSUV

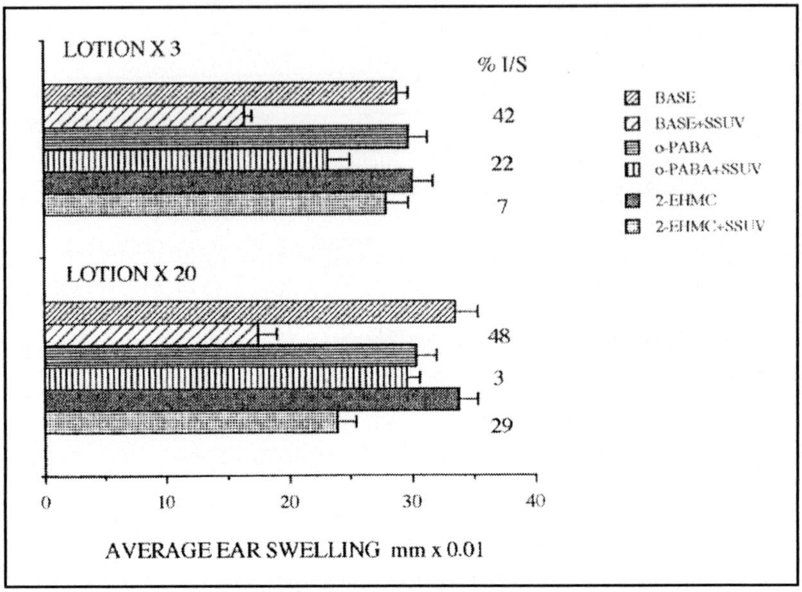

Fig. 9.4. Systemic contact hypersensitivity to oxazolone in mice treated topically on 3 or 20 consecutive days with base lotion or sunscreen lotions, followed 30 min after the last 3 treatments with 1 MED or 6 MED of SSUV, respectively, compared with mice which received only topical lotions but were not irradiated.

exposure, chronic application of 2-EHMC was observed to allow 29% suppression of CHS by SSUV exposure.

This interesting alteration in the photoimmunoprotection by o-PABA and 2-EHMC upon chronic application was supported when exogenous *cis*-urocanic acid was the immunosuppressive agent (Fig. 9.5). Again the chronic application of the base lotion had no effect on the *cis*-urocanic acid-suppression of CHS compared with acute application (35% and 30% suppression, respectively). However, chronic o-PABA application totally abrogated the suppression of CHS from 31% after 3 sunscreen applications, to -8%. With chronic 2-EHMC application, a small degree of suppression of CHS (10%) was observed to result from *cis*-urocanic acid, although this was not statistically significant. Nevertheless, it differed from the total lack of suppression observed after acute application before *cis*-urocanic acid.

Summary

Thus we had observed a total reversal of the influences of o-PABA and probably also 2-EHMC, on SSUV- and *cis*-urocanic acid-

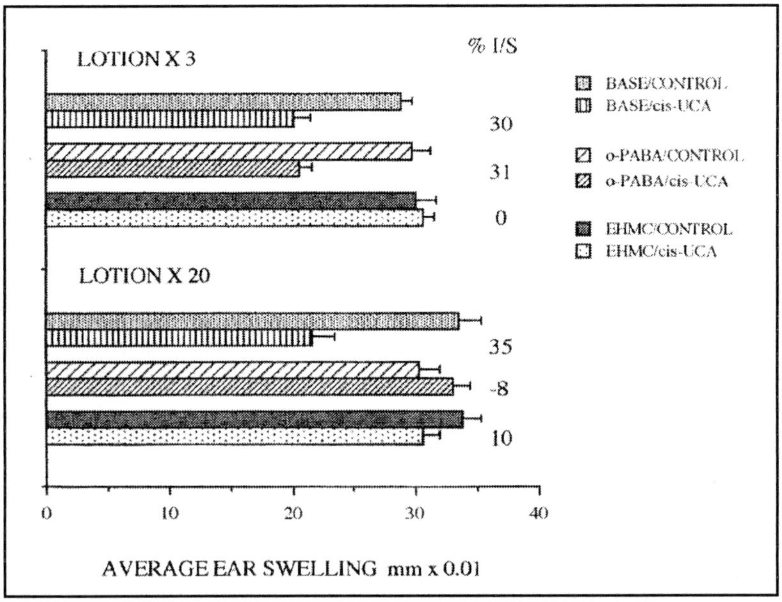

Fig. 9.5. Systemic contact hypersensitivity to oxazolone in mice treated topically on 3 or 20 consecutive days with base lotion or sunscreen lotions, followed 60 min after the last 3 treatments with 200 mg photoisomerized urocanic acid lotion (containing 51% *trans*-, 49% *cis*- UCA), compared with mice which received only topical lotions but not urocanic acid.

induced immunosuppression if the sunscreens were applied repeatedly to the skin. The immunological consequences of irradiation through chronically sunscreen-treated skin are aligned with the photocarcinogenic risk. It is currently not known how this phenomenon comes about. Perhaps o-PABA accumulates gradually in some cutaneous stratum after multiple daily applications, and this pool of o-PABA might obstruct access out of the skin for isomerized urocanic acid. It is more difficult to speculate on the fate of urocanic acid in chronically 2-EHMC-treated skin, however, since 3 applications of 2-EHMC inactivated the *cis* isomer (Fig. 9.3) but 20 applications certainly did not, and might even be reactivating its suppressive function (Fig. 9.5). Further studies are needed to extend these results. However, some obvious conclusions can be drawn: first, that measurement of immune function by acute tests does not necessarily apply in chronically sunscreen-irradiated mice; second, that interactions thus far undescribed, between UVB-absorbers and immunologically regulatory cutaneous molecules appear to occur and may become relevant when daily sunscreen usage is prolonged.

Conclusions

Thus the disparate effects of these two UVB-absorbers in protection from UV-induced immunosuppression appear to be a result of specific inactivation of *cis*-urocanic acid. Acute UV exposures through topical sunscreens have not necessarily reflected the immune status alterations in mice chronically irradiated through suncreens, and the interaction of o-PABA and 2-EHMC with *cis*-urocanic acid varied with the number of daily sunscreen treatments, the effects of acute applications being reversed when the sunscreens were chronically applied. However, the degree of immunosuppression caused by chronic irradiation through sunscreens paralleled the severity of photocarcinogenesis, although not necessarily the protection from acute irradiation-induced erythema/edema.

It is apparent that much remains undefined concerning the interactions of these UVB-absorbers with the skin when they are in constant daily use. Currently o-PABA is in disfavor as a UVB-absorber, mainly because of its photosensitizing properties,[54,55] but its vastly superior anti-photocarcinogenic action suggests that o-PABA derivatives could be developed as future sunscreen ingredients.

Acknowledgments

The excellent research assistance of Caroline Kerr, Meira Bosnic and Christa Boehm-Wilcox, and careful animal husbandry by Lyn Blyth, are gratefully acknowledged. These studies have been supported financially by the N.S.W. State Cancer Council and the University of Sydney Cancer Research Fund.

References

1. Stenbeck KD, Balanda KP, Williams MJ et al. Patterns of treated nonmelanoma skin cancer in Queensland—the region with the highest incidence rates in the world. Med J Aust 1990; 153:511-515.
2. MacLennan R, Green AC, McLeod GR et al. Increasing incidence of cutaneous melanoma in Queensland, Australia. J Natl Cancer Inst 1992; 84:1427-1432.
3. Burton RC, Coates MS, Hersey P et al. An analysis of a melanoma epidemic. Int J Cancer 1993; 55:765-770.
4. Koh HK, Geller AC, Miller DR et al. Prevention and early detection strategies for melanoma and skin cancer—current status (review). Arch Dermatol 1996; 132:436-443.
5. Kinlen LF, Sheil AGR, Peto J et al. Collaborative United Kingdom-Australian study of cancer in patients treated with immunosuppressive drugs. Br J Med 1979; II:1461-1466.

6. O'Dell BL, Jesser RT, Becker LE et al. Diminished immune response in sun-damaged skin. Arch Dermatol 1980; 116:559-561.
7. Boyle J, Mackie RM, Briggs JD et al. Cancer, warts and sunshine in renal transplant patients: a case control study. Lancet 1984; I:702-705.
8. Yoshikawa T, Rae V, Bruins-Slot W et al. Susceptibility to effects of UV radiation on induction of contact hypersensitivity as a risk factor for skin cancer. J Invest Dermatol 1990; 95:530-536.
9. Sheil AG, Disney AP, Mathew TG et al. Malignancy following renal transplantation. Transplant Proc 1992; 24:1946-1947.
10. Fisher MS, Kripke ML. Systemic alteration induced in mice by UV light irradiation and its relationship to UV carcinogenesis. Proc Natl Acad Sci USA 1977; 74:1688-1692.
11. Fisher MS, Kripke, ML. Suppressor T lymphocytes control the development of primary skin cancers in ultraviolet irradiated mice. Science 1982; 216:1133-1134.
12. Norval M, Simpson TJ, Ross JA. Urocanic acid and immunosuppression. Photochem Photobiol 1989; 50:267-275.
13. Norval M, Gibbs NK, Gilmour J. The role of urocanic acid in UV-induced immunosuppression: recent advances (1992-1994). Photochem Photobiol 1995; 62:209-217.
14. Applegate LA, Ley RD, Alcalay J et al. Identification of the molecular target for the suppression of contact hypersensitivity by ultraviolet radiation. J Exp Med 1989; 170:1117-1131.
15. Kripke ML, Cox PA, Alas LG et al. Pyrimidine dimers in DNA initiate systemic suppression in UV-irradiated mice. Proc Natl Acad Sci USA 1992; 89:7516-7520.
16. Gilmour JW, Norval M, Simpson TJ et al. The role of histamine like receptors in immunosuppression induced by *cis*-urocanic acid. Photodermatol Photoimmunol Photomed 1993; 9:250-254.
17. Jaksic A, Finlay-Jones JJ, Watson C et al. *Cis*-urocanic acid synergizes with histamine for increased PGE2 production by human keratinocytes: link to indomethacin-inhibitable UVB-induced immunosuppression. Photochem Photobiol 1995; 61:303-309.
18. De Fabo EC, Noonan FP. Mechanism of immune suppression by UV radiation in vivo. I. Evidence for the existence of a unique photoreceptor in skin and its role in photoimmunology. J Exp Med 1983; 158:84-98.
19. Bestak R, Halliday GM. Chronic low-dose UVA radiation induced local suppression of contact hypersensitivity, Langerhans cell depletion and suppressor cell activation in C3H/HeJ mice. Photochem Photobiol 1996; 64:969-974.
20. Reeve VE, Boehm-Wilcox C, Bosnic et al. Lack of correlation between suppression of contact hypersensitivity by UV radiation and photoisomerisation of epidermal urocanic acid in the hairless mouse. Photochem Photobiol 1994; 60:268-273.
21. Reeve VE, Bosnic M, Boehm-Wilcox C et al. Differential protection by two sunscreens from UV radiation-induced immunosuppression. J Invest Dermatol 1991; 97:624-628.

22. Reeve VE, Boehm-Wilcox C, Bosnic M et al. Differential photoimmunoprotection by sunscreen ingredients is unrelated to epidermal *cis*-urocanic acid formation in the hairless mouse. J Invest Dermatol 1994; 103:801-806.
23. Walker SL, Young AR. Lack of immunoprotection by UVB sunscreens is not due to sunscreen interaction with the skin. Photochem Photobiol 1995; 88S (abstract).
24. Wolf P, Donawho CK, Kripke ML. Analysis of the protective effect of different sunscreens on ultraviolet radiation-induced local and systemic suppression of contact hypersensitivity and inflammatory responses in mice. J Invest Dermatol 1993; 100:254-259.
25. Wolf P, Yarosh DB, Kripke ML. Effects of sunscreens and a DNA excision repair enzyme on ultraviolet radiation-induced inflammation, immune suppression, and cyclobutane pyrimidine dimer formation in mice. J Invest Dermatol 1993; 101:523-527.
26. Roberts LK, Beasley DG, Learn DB et al. Ultraviolet spectral energy differences affect the ability of sunscreen lotions to prevent ultraviolet radiation-induced immunosuppression. Photochem Photobiol 1996; 63:874-884.
27. Gasparro FP. UV-induced photoproducts of para-aminobenzoic acid. Photodermatol 1985; 2:151-157.
28. Bernstein HI, Quigley WC. The photochemical dimerization of trans-cinnamic acid. J Am Chem Soc 1945; 63:1845-1846.
29. Sutherland JC, Griffin KP. ρ-aminobenzoic acid can sensitize the formation of pyrimidine dimers in DNA: direct chemical evidence. Photochem Photobiol 1984; 40:391-394.
30. Shimoi KL, Nakamura Y, Noru T et al. Methyl cinnamate derivatives enhance UV-induced mutagenesis due to the inhibition of DNA excision repair in *Escherichia coli B/r*. Mutat Res 1985; 146:15-22.
31. Ingouville NA, Martincigh BS, Raynor MW et al. The photochemistry of the sunscreen absorber 2-ethylhexyl-p-methoxycinnamate. Proc. 6th Congress of the European Society for Photobiology, Cambridge U.K., Sept. 1995:70.
32. Seim J, Christensen T, Johnsen B et al. Decomposition of sunscreens by ultraviolet radiation. Proc 6th Congress of European Society for Photobiology, Cambridge U.K., Sept. 1995:69.
33. Martincigh BS. The photosensitisation of thymine dimer by sunscreen absorbers. Photochem Photobiol 1996; 63:16S (abstract).
34. Dunford R, McHugh P, Knowland J. Sunscreen ingredients that attack DNA by producing free radicals on illumination. Photochem Photobiol 1996; 63:16S (abstract).
35. Allan JM, Gossett CJ, Allen SK. Photochemical formation of singlet molecular oxygen ($^1O_2$) in aqueous solutions of several sunscreen active ingredients. Photochem Photobiol 1996; 63:16S (abstract).
36. Flindt-Hansen H, Thune P, Nielsen CJ. Photocarcinogenesis is retarded by a partly photodegraded solution of para-aminobenzoic acid. Photodermatol 1989; 6:263-267.

37. Snyder DS, May M. Ability of PABA to protect mammalian skin from ultraviolet light-induced skin tumors and actinic damage. J Invest Dermatol 1975; 65:543-546.
38. Kligman LH, Akin FJ, Kligman AM. Sunscreens prevent ultraviolet carcinogenesis. J Am Acad Dermatol 1980; 3:30-35.
39. Wulf HC, Poulsen T, Brodthagen H et al. Sunscreens for delay of ultraviolet induction of skin tumors. J Am Acad Dermatol 1982; 7:194-202.
40. Gallagher CH, Greenoak GE, Reeve VE et al. UV carcinogenesis in the hairless mouse skin. Influence of the sunscreen 2-ethylhexyl-p-methoxycinnamate. Aust J Exp Biol Med Sci 1984; 62:577-588.
41. Reeve VE, Greenoak GE, Gallagher CH et al. Effect of immunosuppressive agents and sunscreens on UV carcinogenesis in the hairless mouse. Aust J Exp Biol Med Sci 1985; 63:655-665.
42. Flindt-Hansen H, Thune P, Larsen TE. The inhibiting effect of PABA on photocarcinogenesis. Arch Dermatol Res 1990; 282:38-41.
43. Reeve VE, Kerr CA. Comparative protection from photocarcinogenesis by two UVB-absorbers matched for SPF in hairless mice. Photochem Photobiol 1996; 63:17S-18S (abstract).
44. Gibbs NK, Norval M, Traynor NJ et al. Action spectra for the *trans* to *cis* photoisomerisation of urocanic acid in vitro and in mouse skin. Photochem Photobiol 1993; 57:584-590.
45. Kammeyer A, Teunissen MBM, Pavel S et al. Photoisomerization spectrum of urocanic acid in human skin and in vitro—effects of simulated solar and artificial ultraviolet radiation. Br J Dermatol 1995; 132:884-891.
46. Reeve VE, Greenoak GE, Canfield PJ et al. Topical urocanic acid enhances UV-induced tumour yield and malignancy in the hairless mouse. Photochem Photobiol 1989; 49:459-464.
47. Katiyar SK, Elmets CA, Agarwal R et al. Protection against ultraviolet-B radiation-induced local and systemic suppression of contact hypersensitivity and edema responses in C3H/HeN mice by green tea polyphenols. Photochem Photobiol 1995; 62:855-861.
48. Cope RB, Bosnic M, Boehm-Wilcox et al. Dietary butter protects from ultraviolet radiation-induced suppression of contact hypersensitivity in the Skh:HR-1 mouse. J Nutr 1996; 126:681-692.
49. Reeve VE, Bosnic M, Boehm-Wilcox C. Dependence of photocarcinogenesis and photoimmunosuppression in the hairless mouse on dietary polyunsaturated fat. Cancer Letts 1996; 108:271-279.
50. Reeve VE, Matheson MJ, Bosnic M et al. The protective effect of indomethacin on photocarcinogenesis in hairless mice. Cancer Letts 1995; 95:213-219.
51. Matheson M, Reeve VE. The effect of the antihistamine cimetidine on ultraviolet radiation-induced tumorigenesis in the hairless mouse. Photochem Photobiol 1991; 53:639-642.
52. Reeve VE, Matheson MJ, Greenoak et al. Effect of dietary lipid on UV light carcinogenesis in the hairless mouse. Photochem Photobiol 1998; 48:689-696.

53. Wang ZY, Agarwal R, Bickers D et al. Protection against ultraviolet B radiation-induced photocarcinogenesis in hairless mice by green tea polyphenols. Carcinogenesis 1991; 12:1527-1530.
54. Kaidbey KH, Kligman AM. Phototoxicity to a sunscreen ingredient, Padimate A. Arch Dermatol 1978; 114:547-549.
55. Thune P. Contact and photocontact allergy to sunscreens. Photodermatol 1984; 1:5-9.

CHAPTER 10

Epilogue: New Perspectives in Sunscreen Photobiology

Francis P. Gasparro

Rationale for This Chapter

While an awareness of the harmful effects of sunlight are more appreciated today than ever before, more and more humans are exposing themselves to greater and greater doses of UV light. For the most part the source of this exposure is the sun, but exposure to artificial sources cannot be ignored. Fortunately public health messages warning about overexposure have become prominent and widespread and hence, over the past 10 years or so there has been an effort to increase sun protection either by recommending covering up or by recommending sunscreen use by people who need (or want) to spend more time outdoors.

Over this same period of time there has been an alarming increase in the incidence of skin cancer.[1] Clearly, greater exposure to sunlight must have some role in this increased incidence of skin cancer. It has also been suggested there might be a link to the increased use of sunscreens.[2] The latter is often dismissed because of the short time—10 to 15 years—that sunscreens have been in vogue when compared to the expected lag time for the development of skin cancer (20-30 years). However, this reasoning ignores the factor that sunlight, in addition to being an initiator of cancer, is also a promoter.[3] Furthermore, there are other ways sun exposure can affect tumor progression (see below). The use of sunscreens would be expected to attenuate the promotional steps and thus it may not be too much to expect sunscreens to reduce the incidence of skin cancer. This concern led to two epidemiological studies in Europe which concluded there could be a link between sunscreen use and skin cancer.[4,5]

Sunscreen Photobiology: Molecular, Cellular and Physiological Aspects, edited by Francis P. Gasparro.

In the foregoing chapters the effects of UV on sunscreens and the ability of sunscreens to prevent erythema and skin immune suppression have been reviewed. Although there can be discussion concerning the finer points, it is clear that the existing assays that are used to test sunscreens are appropriate for the evaluation of short-term efficacy. They are valuable physiological measurements that elucidate immediate early physiological responses in skin after exposure to radiation. These observations appear to be supported at the molecular level by evidence for reduced levels of DNA photoproducts in skin protected with sunscreens.[6] Yet we do not know if the reduction in DNA damage alone is sufficient to reduce the incidence of skin cancer. The amount of DNA damage induced by suberythemal levels of UV could lead to as of yet undefined epigenetic events.

Two issues are examined in this chapter. The first concerns whether there are other kinds of markers whose expression may be induced by cellular damage (and not necessarily by DNA damage). The second concerns the possibility that there may be other endogenous chromophores in skin that lead to deleterious effects when chronically exposed to solar radiation.

Molecular Markers

Among these markers are molecules like NFκB, AP-1 and p53. These molecules may be involved in some of the well-characterized events that follow sun exposure: DNA repair, sunburn and suntanning. But they may have other, more subtle, effects occurring over a longer time period. NFκB is the prototypical transcription factor. It is released from a cytoplasmic complex when a cell is damaged. The transcription factor AP-1, a combination of c-*fos* and c-*jun*, is another example of a UV-inducible protein which regulates the expression of specific genes. p53 has been called the guardian of the genome. It is a very important molecule that determines the fate of a cell once it is damaged.

NFκB could be called a cellular sentinel. It is preformed, sitting in the complex with I-kappa B (IκB) in the cytoplasm of a cell.[7] A cell that suffers damage—and there has been some controversy about what kind of damage—but it may be either nuclear damage or cell membrane damage, sends a signal that leads to the phosphorylation of IκB, inducing a conformational change and the release of NFκB which then translocates to the nucleus, binds to corresponding binding sites and upregulates the expression of certain genes.

Another kind of event likely to occur in damaged skin is the induction of transcription factors like AP-1 and ERG-1.[8] These are early response genes which play a role in determining the fate of the cell. These molecules can be induced with suberythemal doses of UV light and hence may be active in skin with no evidence of sunburn.

Much has been written about p53[9] but only a one of these reports dealt with its activity in sunscreen protected skin (SPS).[10] One of the important functions of p53 is its ability to arrest cell cycle progression, so that a cell with severe damage does not go on to replicate. In essence it makes a decision, if we can give it a humanlike quality, that a cell should not replicate, and it pushes the cell off towards the apoptotic pathway, eliminating the cell all together, and eliminating it as a future problem.

The paramount question that needs to be answered is—Do sunscreens prevent or minimize the induction of these molecules (or others like them)? If not, what is the wavelength dependence for their induction? It is very important to know the action spectrum because sunscreens are not absolute blocks. The altered spectrum that penetrates SPS may not cause sunburn, but that is not to say it cannot have other effects (perhaps deleterious and perhaps delayed). Regarding action spectra it must also be appreciated that nothing special happens to photons on going from 399 nm to 400 nm to 401 nm. The objective definition of the boundary between UVA and visible based on what the human eye 'detects' becomes an arbitrary one when applied to skin photobiology. Violet and blue photons are used to treat neonatal jaundice[11] and it is interesting to note that recent studies by Dick Setlow of Brookhaven have illustrated their ability to induce melanoma in his fish model.[12] Thus, action spectra need to be extended into the short wavelength region of the visible spectrum. A novel mechanism for skin photosensitization to longer wavelengths of ultraviolet light (beyond ~360 nm) is presented below.

Induction

Campbell et al showed a strong wavelength dependence for p53 induction in unprotected skin using equally erythemogenic doses or UVA, B, and C.[13] There was also a difference in the pattern of distribution in skin. With UVC, the staining is confined to the upper epidermis, which is not surprising since that is probably as far as UVC penetrates. The staining in skin exposed to UVB and UVA also correlated with the depths to which they penetrate human skin. UVB penetrates the epidermis and UVA reaches the basal layer. At this time no attempts have been made to determine if short wavelength visible light induces p53.

In another study looking at the effect of sunscreens on p53 and p21, a downstream effector of p53, the p53 is expressed in the epidermis after radiation—in skin without sunscreen while p21 was expressed both in the dermis and epidermis.[10] Natural pigmentation appeared to block the induction of p53 and p21. Furthermore, topically applied sunscreen reduced or eliminated the expression of these markers. There was a delayed time course for the expression of these molecules, up to 48 hours, which indicates that something was occurring at the molecular level. A shift in the pattern of staining with UVA from the superbasal layer to the basal layer over that same time period was also observed. Up to 85 to 90% of all skin specimens, with the exception of those taken from parts of the body not usually exposed to sunlight, showed at least a slight p53 reactive pattern. It appears that low doses of sunlight could prime the skin for increased expression of p53 and p21. These observations may impact on how we view people's lifestyle—whether they use a sunscreen all the time or just some of the time or whether it is due to forgetfulness or just not wanting to bother with it. Nevertheless, there are health consequences that may result from different kinds of photobiological events which will depend on whether the skin is protected (partially or fully from erythema).

It was also interesting to note that sunburn cells—either with or without sunscreens—were not observed; this is different from the kinds of observations that are seen in mice, which raises another point about the validity of making leaps from mice to men.[10] Mice are certainly convenient systems for doing large studies with lots of controls, but we have to be careful about extension of the interpretation of the results to human skin.

In the studies described by Pontén et al, the sunscreen did prevent p53 induction when 1 MED was delivered, but when sunscreens are used in real life, people stay out in the sun for a longer time and are exposed to more than 1 MED.[10] It may be unlikely that people stay out in the sun 15 times as long, but it does raise this question—What would have happened to p53 expression if the sunscreen-protected skin received 5 or 10 MEDs? This important question needs to be considered when evaluating the usefulness of various markers.

There are two very important aspects of these studies. First, they were performed in human skin. As attractive as animal studies are from the point of view of controlling all experimental parameters, they must be interpreted cautiously before being applied to humans. Second, p53 plays a role in determining the fate of a cell and hence a molecular event is measured rather than some endpoint (erythema, tanning or immune suppression). There is one drawback in the above cited studies—sunlight was not used. Convenient experiments can be performed with solar simulated light, but the best light for the evaluation of impact on human skin is that from the sun. Another shortcoming of these studies is that the potential effects of short wavelength visible light were not considered.

Epigenetic Effects

There may be other ways the critical activity of p53 may be attenuated (other than mutated). p53 regulates the expression of (1) *mdm-2*—a protein that also interacts with p53 itself and blocks its transactivating effects; (2) *gadd45*—a protein which binds to PCNA and is involved in DNA replication and excision repair; and (3) p21(Cip1/Waf1) - a protein which inhibits cell division through a process involving interactions with cyclin dependent kinases. Thus, *mdm-2* and *gadd45* link p53 to inducible DNA repair processes and p21 to the G_1 checkpoint system.[14a] It would not be surprising to find molecules such as these playing a role in skin exposed to UV radiation—sunscreen protected or not. *bcl-2* is an antagonist of p53 activity. It is unique in at least two respects: First, the protein is located in mitochondria; Second, it interferes with apoptosis by an as of yet undetermined mechanism. The end result of its presence in cells is the promotion of cell survival (as opposed to stimulating cell proliferation). Hence increased *bcl-2* levels can confer resistance to apoptosis in cells exposed to UV radiation.[14b]

In conclusion, the physiological assays, looking at prevention of erythema or immune suppression, may be adequate for initial screening of potential sunscreen ingredients, that is, to determine their physiological efficacy in the prevention of sunburn and other short-term deleterious effects. But for long-term efficacy and safety, new molecular assays are required that reflect the potential fate of sunscreen-exposed cells in human skin to long-term effects. In terms of applying the techniques of modern molecular biology to the study of sunscreens, it should be noted that these kinds of studies would answer questions about which molecular events are occurring in human skin. This would be in contrast to the photo-Ames test which would best serve as a preliminary screening assay.[15] The ultimate test for efficacy and long-term safety should be performed in human cells. Perhaps the technique for assaying mutagenicity in the *hprt* gene of human T cells could be adapted to skin cells.[16] Finally, there are suggestions that there may be chromophores in skin that absorb longer wavelengths of ultraviolet light and/or visible light.[17] With naturally occurring endogenous molecules in the skin or perhaps sunscreens that absorb some of these wavelengths of light, subtle effects could be missed if we just focus on ultraviolet light.

Iron Deposition in Skin Chronically Exposed to UV

Chronic exposure of skin to ultraviolet light leads to a significant increase in skin iron content.[18] Several investigators have related iron in skin exposed to ultraviolet light to the generation of singlet oxygen and subsequent deleterious effects of UVA on lipids and proteins.[19] The topical application of iron-chelating agents like 2- fluridioxime (FDO), can dramatically reduce these effects.[20] In a study in hairless mice, the application of either FDO or padimate-O delayed skin wrinkling after UVB exposure by a factor of 2-3. When FDO and padimate-O were applied simultaneously, they acted synergistically to delay skin wrinkling by nearly an order of magnitude. Aubailly et al has shown the photoreduction of ferritin by UVA light leads to the peroxidation of model protein solutions.[21] The UVA doses employed are in the range that would produce a mild response in human skin and their conclusion that "..skin ferritin, among others, cannot be neglected as a photochemically active chromophore of cutaneous cells.." points towards a new agent that needs to be considered as having a possible role in the induction of deleterious effects in human skin chronically exposed to sunlight (whether unfiltered or SPS-filtered).

Most sunscreens only partially screen skin from UVA. Granted that the filtered wavelengths are in the shorter, more potent range of UVA (<350 nm) which are those that have been shown to have direct effects on DNA. However, when the possibility of iron-sensitized photoreactions are considered, the range of wavelengths that are potentially activating probably extend further towards the visible region of the solar spectrum. Given the large amounts of longer wavelength UVA and shorter wavelength visible photons that sunscreen-protected skin may receive, these other photons may have potent effects.

Thus far studies on iron deposition and sensitized reactions in skin have employed UV light (UVA and UVB). The action spectrum for iron deposition needs to be determined. If iron can find its way into skin one way or another and then become photoactivatable by ultraviolet, and perhaps the shorter wavelengths of visible light, it may lead to cellular effects which induce molecules like NFκB and AP-1 or even p-53. Visible light responses are not uncommon in photobiology. For example, there is a well-defined blue light response in plants.[22] In addition, visible light therapy is used in the treatment of neonatal jaundice.[12] Christensen et al showed that the decomposition of bilirubin could be accomplished using wavelength of light as far into the visible as 530 nm.[23] Thus, it might not be surprising to find that iron deposited in skin chronically exposed to UV light (and perhaps even SPS-filtered UV light) is leading to subtle effects in skin that lead to cellular transformations.

Summary

The issues discussed in this book have needed to be discussed for a long time. There appears to be a surge in interest in studying sunscreens. A Medline search for the occurrence of the term 'sunscreen' in the title or abstract shows a dramatic increase in the number of papers for 1991-1995 compared to the preceding five years. In the most recent period 186 papers were published, for 1986-90, the number was 98 while between 1966 and 1975 there are only 20 citations. Many of these earlier studies were purely physiological, but more and more of them are becoming 'mechanistic' and 'molecular'. This is a healthy trend that should lead to significant improvements in our understanding of what makes a good sunscreen and, more importantly, contribute to our understanding of the effects of sunlight of all wavelengths on skin biology.

Notes Added in Proof

Two papers relevant to the topics discussed in this chapter have appeared since its submission.

Luther et al describe a study on the use of sunscreens in children and the occurrence of melanocytic nevi. The main goal was to examine nevus development in children and related factors such as pigmentation phenotype and the role of sun exposure. Their not too surprising results were that the development of nevi was related to the two cited factors. However, they also found a positive relation between the regular use of sunscreens and a high number of acquired nevi. The authors stated that "Children who use sunscreens regularly tend to have more cumulative sun exposure..." They attributed this increase to inadequate application and variable reapplication. Another interpretation is that the wavelengths of light responsible for nevus development are not the same wavelengths blocked by the sunscreen (i.e. not the burning UVB rays!). Hence the increased number of nevi may have arisen as a result of light that penetrated the sunscreen protected skin. (Luther H, Altmeyer P, Garbe C et al. Increased melanocytic nevus counts in children during 5 years of follow-up and analysis of associated factors. Arch Dermatol 1996; 132:1473-1478.

Ananthaswamy et al described the ability of two sunscreens to inhibit the induction of p53 mutations in UV-irradiated mouse skin. In order to avoid the tedious task of examining all 11 exons of p53, these authors selected a site that is mutated in 27% of UV-induced skins tumors in mice for sequence analysis. They showed that the application of sunscreens before each irradiation nearly abolished the occurrence of p53 mutations at the selected site. Artificial light emitting only a portion of the solar spectrum were employed in these studies, which means that these mice were not exposed to the high doses of longer wavelength UVA and shorter wavelength visible light that is contained in the solar spectrum (and which, for example, may have contributed to the development of the nevi in the study cited above). Nonetheless, this is an important study because it examined the effects of sunscreens on a molecule that influences the fate of a cell. More studies like this need to be done in human skin and with a physiologically relevant light source. (Ananthaswamy HN, Loughlin SM, Cox P et al. Sunlight and skin cancer: inhibition of p53 mutations in UV-irradiated mouse skin by sunscreens. Nature Medicine 1997; 3:51-514.)

SPF or SBF

The results of the two papers cited above emphasize the need to consider the impact of all photons that fall on human skin and not just those that cause sunburn. As mentioned many times in this book, sunscreens are very good at protecting skin from sunburn. We have seen, however, that there are other effects in skin. We must be careful not to confuse the consumer about what a sunscreen can and cannot do until these latter effects are understood. Perhaps we should be referring to the SunBurn protection factor (SBF) and not the Sun Protection Factor (SPF).

References

1. Miller DL, Weinstock MA. Nonmelanoma skin cancer in the United States: incidence. J Am Acad Dermatol 1994; 30:774-778.
2. Garland CF, Garland FC, Gorham ED. Rising trends in melanoma. An hypothesis concerning sunscreen effectiveness. Ann Epidemiol 1993; 3:103-110.
3. Kripke ML, Wolf P. Effect of sunscreens on UV radiation-induced enhancement of melanoma growth in mice. J Natl Cancer Inst 1994; 86:798-801.
4. Autier P, Dore JF, Schifflers E et al. Melanoma and use of sunscreens: an EORTC case-control study in Germany, Belgium and France: The EORTC Melanoma Cooperative Group. Int J Cancer 1995; 61:749-755.
5. Westerdahl J, Olsson H, Masback A et al. Is the use of sunscreens a risk factor for malignant melanoma. Melanoma Res 1995; 5:59-65.

6a. Wolf P, Yarosh DB, Kripke ML. Effects of sunscreens and a DNA excision repair enzyme on ultraviolet radiation-induced inflammation. Immune suppression and cyclobutane pyrimidine dimer formation in mice. J Invest Dermatol 1993; 101:523-527.

6b. van Praag MC, Roza L, Boom BW, Out-Luijting. Determination of the photoprotective efficacy of a topical sunscreen against UVB-induced DNA damage in human epidermis, J Photochem Photobiol 1993; B19:129-134.

7. Pahl HL, Baeuerle PA. Oxygen and the control of gene expression. Bioessays 1994; 16:497-502.
8. Holbrook NJ, Fornace AJ. Response to adversity: molecular control of gene activation following genotoxic stress. New Biologist 1991; 3:825-833
9. Wang XW, Harris CC. TP53 tumor suppressor gene: clues to molecular carcinogenesis and cancer therapy. Cancer Surveys 1990; 28:169-196.
10. Pontén F, Berne B, Ren Z-P, Nistér M, Pontén J. Ultraviolet light induces expression of p53 and p21 in human skin: effect of sunscreen and constitutive p21 expression in skin appendages. J Invest Dermatol 1995; 105:402-406.
11. Ennever JF, Know I, Denne SC, Speck WT. Phototherapy for neonatal jaundice: in vivo clearance of of bilirubin photo-products. Pediatr Res 1985; 19:205-208.
12. Setlow RB, Grist E, Thompson K, Woodhead AD. Wavelengths effective in induction of malignant melanoma. Proc Natl Acad Sci USA 1993; 90:6666-6670.
13. Campbell C, Quinn AG, Angus B, Farr PM, Rees JL. Wavelength specific patterns of p53 induction in human skin following exposure to UV radiations. Cancer Res 1993; 53:2697-2699.

14a. Finlay CA. The *mdm-2* oncogene can overcome wild-type p53 suppression of transformed cell growth. Mol Cell Biology 1993; 13:301-306.
14b. Miyashita T, Reed JC. *bcl-2* gene transfer increases relative resistance of S49.1 and WEHI7.2 lymphoid cells to cell death and DNA fragmentation induced by glucocorticoids and multiple therapeutic drugs. Cancer Res 1992; 52:5407-5411.
15. Chetelat A, Albertini S, Dresp JH, Strobel R, Gocke E. Photomutagenesis test development I. 8-Methoxypsoralen, chlorpromazine and sunscreen compounds. Mutat Res 1993; 292:241-250.
16. Yang J-L, Maher VM, McCormick JJ. Amplification and direct nucleotide sequencing of cDNA from the lysate of low numbers of diploid human cells. Gene 1989; 83:347-354.
17. Jurkiewicz BA, Buettner GR. UV radiation-induced free radical formation in skin. Photochem Photobiol 1996; 63S:105-106.
18. Bissett DL, Chatterjee R, Hannon DP. Chronic ultraviolet radiation-induced increase in skin iron and the photoprotective effect of topically applied iron chelators. Photochem Photobiol 1991; 54:215-223.
19. Vile GF, Tyrrell RM. UVA radiation-induced oxidative damage to lipids and proteins in vitro and inhuman skin fibroblasts is dependent on iron and singlet oxygen. Free Radic Biol Med 1995; 18:721-730.
20. Bissett DL, McBride JF. Synergistic topical protection by a combination of the iron chelator 2-furildioxime and sunscreen. J Am Acad Dermatol 1996; 35:546-549.
21. Aubailly M, Salmon S, Haigle J et al. Peroxidation of model lipoprotein solutions sensitized by photoreduction of ferritin by 365 nm radiation. J Photochem Photobiol 1994; B26:185-191.
22. Senger H. Blue Light Effects in Biological Systems. Springer-Verlag:Berlin, 1984.
23. Christensen T, Kinn G, Granli T, Amundsen I. Cells, bilirubin and light: formation of bilirubin photoproducts and cellular damage at defined wavelengths. Acta Pediatr 1994; 83:7-12.

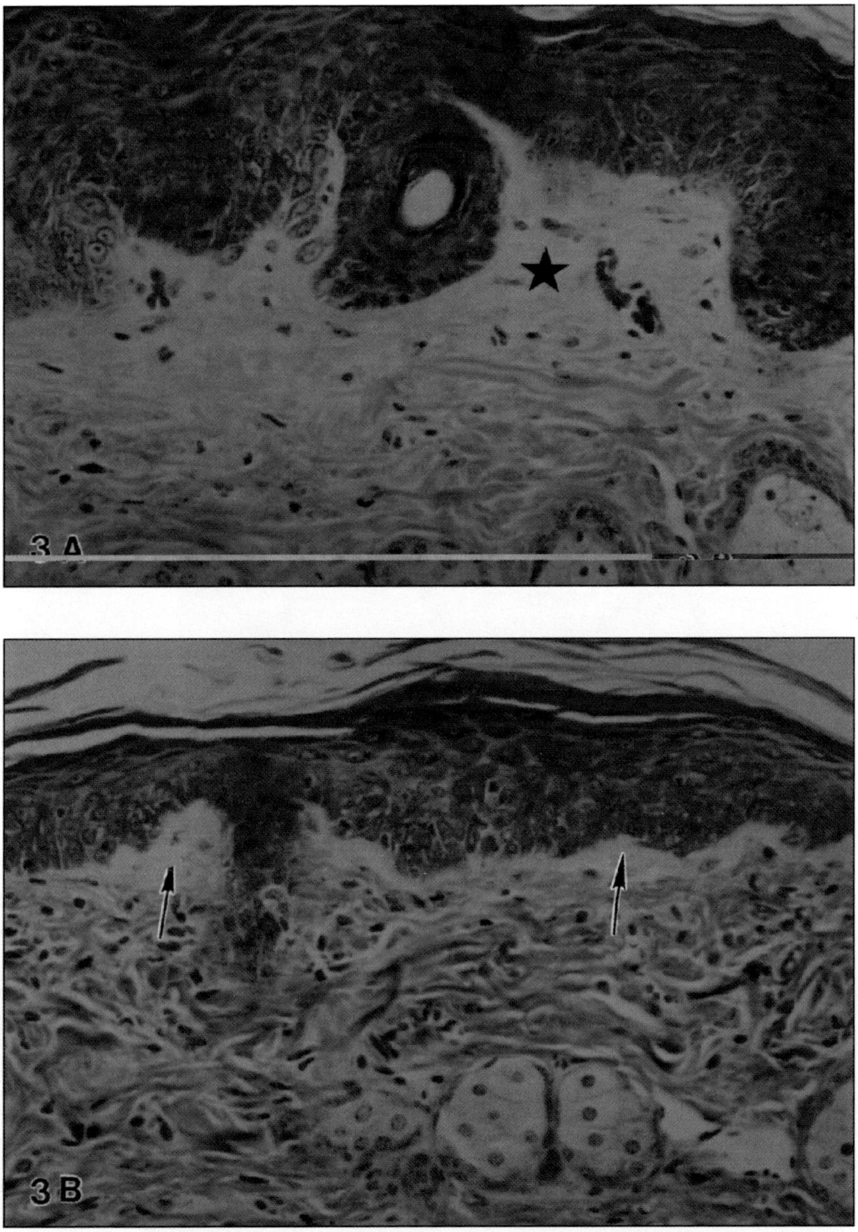

Fig. 5.3. 16 MPD: Collagen. Masson's Tri-Chrome Stain. Magnification x295, (a) SPF 7 Sunscreen. Loss of affinity for the stain is marked but is focal (*). (b) SPF 18 Sunscreen. Collagen staining is normal. The pale subepidermal deposits (→) are shown to be glycosaminoglycans when Mowry's stain is used.

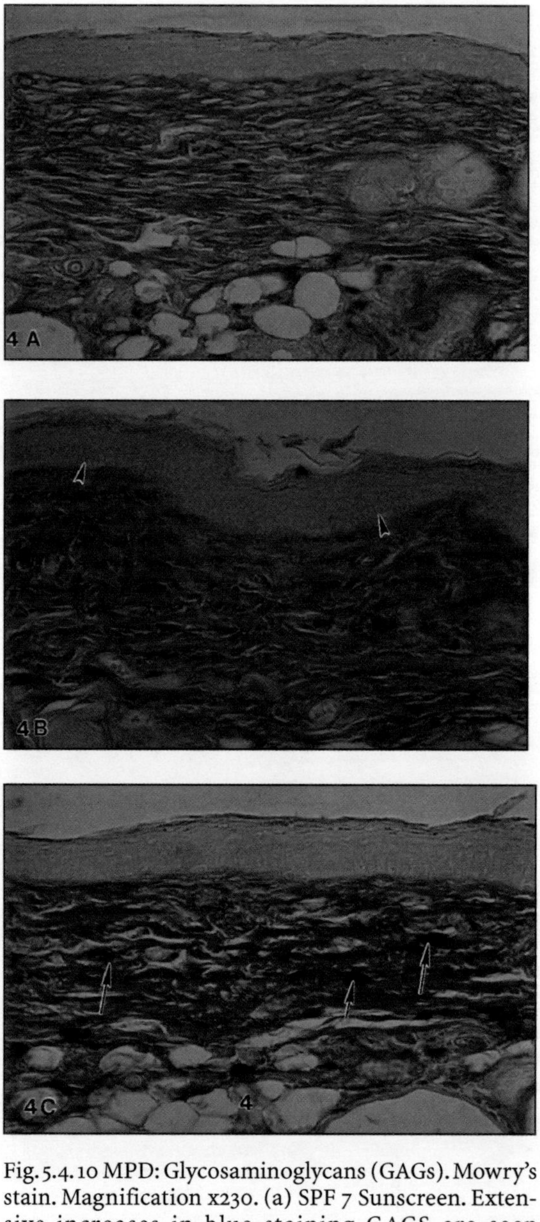

Fig. 5.4.10 MPD: Glycosaminoglycans (GAGs). Mowry's stain. Magnification x230. (a) SPF 7 Sunscreen. Extensive increases in blue staining GAGS are seen subepidermally. (b) SPF 16 Sunscreen. An area with barely discernible dermal GAGs but intercellular epidermal GAGs are present (▶). (c) SPF 18 Sunscreen. GAGS are not increased. The only blue staining dermal features are mast cells (➔).

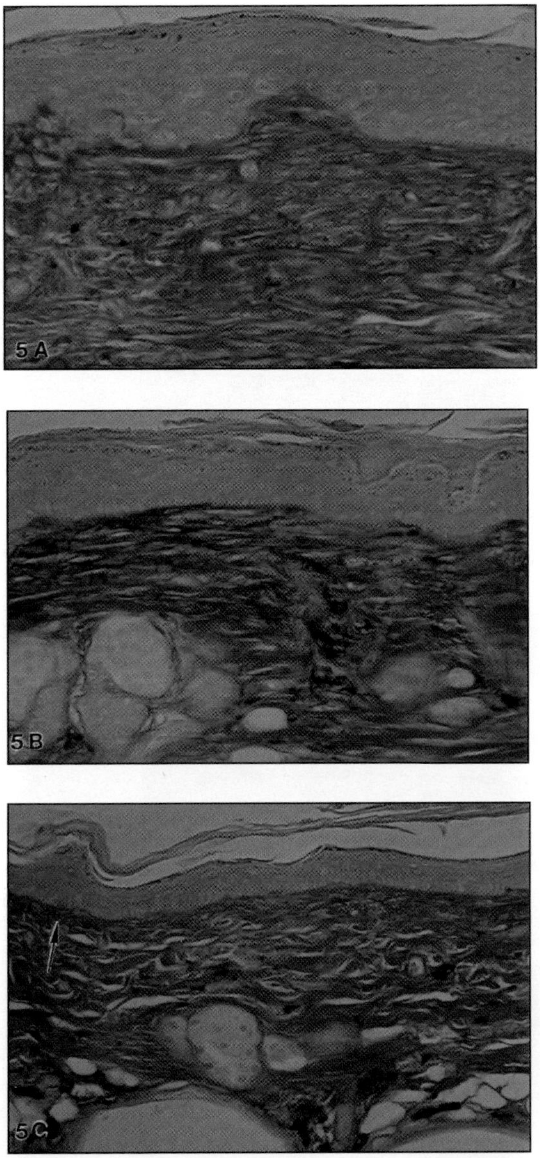

Fig. 5.5. 16 MPD: Glycosaminoglycans (GAGs). Mowry's Stain.Magnification x230. (a) SPF 7 Sunscreen. Blue staining GAGS are considerably increased compared to 10 MPDs. (b). SPF 16 Sunscreen. GAGs are as abundant as with SPF 7 sunscreen. (c) SPF 18 Sunscreen. GAGs are mildly increased with foci of dense deposits at the dermal-epidermal junction (➔). All figures are reprinted from reference 23 with permission from coauthors Agin and Sayre.

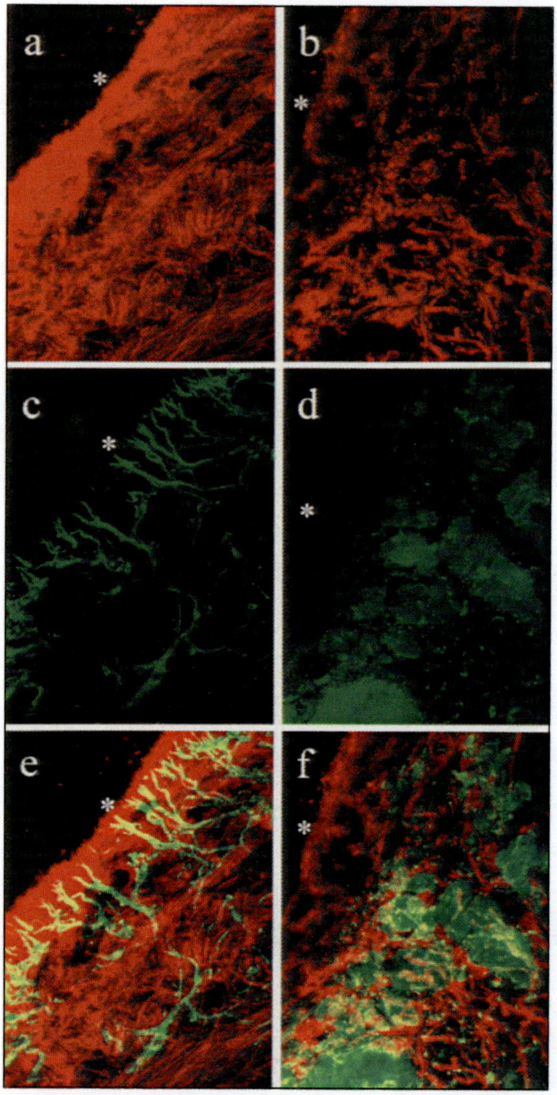

Fig. 6.1. Laser Scanning Confocal Microscopy of Normal and Sun-damaged skin from the same individual. (a) Elastic fibers normally form a delicate meshwork of fibers which extends perpendicularly from the dermal-epidermal junction in sun-protected skin. (b) Chronically photodamaged skin contains numerous, very large, clumped elastic fibers in the superficial dermis replacing the normal architecture. (c) Sun-protected skin also reveals strongly staining collagen just beneath the dermal-epidermal junction. Below this is a dense network of fibers. (d) Sun-damaged skin also contains a band of strongly staining collagen. However, the collagen below this area is stringy and much less dense than sun-protected skin. (e) Double-labelled images show the interrelationship between collagen and elastic fibers. Collagen is red and elastin appears green or yellow. Sun-protected skin demonstrates an orderly arrangement of fibers. (f) In contrast with protected skin, sun-damaged skin reveals a haphazard arrangement of the fibers. Asterisk indicates dermal-epidermal junction in e and f *Bar*, 10 μm. Reprinted with permission from reference 39.

Index

A

α-tocopherol, 111
acquired immunodeficiency syndrome (AIDS), 103
actinic keratoses, 63, 127, 145
action spectrum, 1-8, 64, 71, 105, 111, 116, 136, 140-142, 158, 179, 183
action spectrum for affecting DNA, 6
action spectrum for skin cancer, 4, 6
action spectrum for the immunosuppression, 4
Ames tests, 51, 54
antigen presentation, 100, 103-104
average intensity, 2, 3
avobenzone (Parsol 1789), 72

B

band gap, 28, 56-57
basal, 48, 100, 180
basal layer, 180
Benzophenone-3, 17, 36-37, 64, 106-107, 109, 110, 113-115, 159-160
Benzophenone-8, 15, 36-37
blue light response, 183
Borrelia burgdorferi, 103
broad spectrum, 69-70, 81, 118
butyl-methoxydibenzoyl-methane, 113

C

Candida albicans, 102
cell membrane damage, 118, 178
cellular atypia, 73-74
cellular sentinel, 178
chemical absorbers, 13, 19, 26-28, 37, 39
chemical UV filters, 106-107, 110
chloramphenicol acetyltransferase (CAT), 90
chondroitin sulfate, 89
cis-trans photoisomerization, 24
Cockayne's syndrome, 101
collagen damage, 75
collagenase, 85
connective tissue, 69-70, 79, 96
contact hypersensitivity, 100, 104, 109, 115, 128, 158, 159, 160–161, 165-166, 169-171
cutaneous melanoma, 101-102
[2 + 2] cyclobutane adduct, 34
cyclobutane, 6, 34, 37-38, 47, 52, 67, 105, 110, 116-118
cytokines, 85-86, 91, 100, 119
cytosine, 37
cytotoxicity, 41

D

decorin, 88-89
delayed type hypersensitivity, 102, 128
Dermal damage, 69
desmosines, 88
DMSO, 54-55
DNA as a target, 5, 7
DNA base damage, 37
DNA damage, 7, 32, 47, 49, 55, 57, 59, 100, 101, 105, 117-118, 127, 145, 158, 178
DNA repair, 51, 100-101, 117-119, 178, 181

E

2-EHMC, 159
2-ethylhexyl-para-methoxycinnamate, 34, 35
effectiveness of sunscreens, 66, 106
elastic, 71, 73-75, 80, 85-89
elastic fiber hyperplasia, 71, 74-75
elasticity, 87
electron spin resonance spectroscopy, 52
epidermal hyperplasia, 70, 73-74, 117
epigenetic events, 178
erythema, 1, 7-8, 13, 37, 63-66, 70-73, 75, 77, 79-81, 99, 110, 115, 127, 130-133, 136, 138, 140, 145, 149-150, 158-160, 163-166, 172, 178-181
erythemogenic UV energy, 130-131
ethyl 4-dimethylaminobenzoate, 50
excited state, 20-25, 36, 55
extracellular matrix, 86, 89, 95

F

2- fluridioxime (FDO), 182
Fenton-type reactions, 31
ferritin, 182
fibrillin, 87-88
fibroblasts, 85-86, 90-93
fish model, 179
fluence, 1-5

fluorescence, 22, 28-29, 42
free radicals, 23, 29-32, 41-42, 49, 50, 52-53, 55

G

gadd45, 181
glycosaminoglycans, 73, 76-79, 86-89

H

Haber-Weiss, 26, 31
hapten, 104
hematoxylin and eosin, 73
herpes labialis, 65, 127
herpes simplex virus, 102
hprt gene, 182
human erythema, 8, 132
hydrogen peroxide, 26, 31
hydroxyl radical, 26, 31-32, 54-57

I

IL-10, 105
illumination source, 32
immune protection, 106-108, 110-112, 116-118, 127-129, 132-134, 136-150
immune suppression, 99-102, 104-107, 110-112, 116-119, 138, 145-148, 178, 181
immunodeficiency virus, 103
immunoglobulin, 113
immunohistochemical staining, 88
immunologic tolerance, 100
immunoprotection, 158, 163, 168, 170
immunosuppression, 4, 47, 65, 70, 127-129, 133-134, 136-144, 146, 148-150, 158, 161, 164-165, 168, 171-172
immunosuppression dose, 129
immunosuppression effectiveness spectra, 134, 141
indirect effects on DNA, 8
induced papillomas, 3
induced sarcomas, 3
infectious diseases, 99, 102, 104-105
inflammation, 80, 107, 110-112, 116-118, 150
initiator, 50, 177
internal conversion, 22, 28
intersystem crossing, 22-23, 36
iron deposition, 182-183
iron-chelating agents, 182
isomerization, 4-5, 6, 34, 35, 105, 163, 166

L

Langerhans cells, 103-105, 107, 110, 113, 128, 138
Leishmania major, 103
lipids, 1, 30, 49, 182
Listeria monocytogenes, 104
local CH, 128, 137-138

M

8-methoxypsoralen (8-MOP), 94
M. lepraemurium, 103
Masson's trichrome, 73
mdm-2, 181
melanoma induction in fish, 8
methoxycinnamate, 14-15, 17, 34-36, 65, 72, 106, 109, 115, 159
Michler's ketone, 50
micronized particles, 57
minimal erythema dose, 1, 7, 65-66, 70, 115, 130
minimal photoaging dose (MPD, 71
mitochondria, 181
mixed leukocyte reaction, 113
Mowry's colloidal iron, 73
mRNAs, 90
mutagenesis, 6, 118
mutagenicity in human cells, 8
mutagenicity tests, 50, 55
Mycobacterium bovis bacillus Calmette-Guérin (BCG), 103

N

natural killer cell activity, 101, 113
neonatal jaundice, 179, 183
NFkB, AP-1 and p53, 178
N,N'-dimethylethylenediamine, 53, 54
non-Hodgkin's lymphoma, 102
nonsolar UV spectra, 148
nuclear damage, 178
nucleotide excision repair, 6

O

Octocrylene, 17, 36
octyl methoxycinnamate, 17, 65, 72
octyl salicylate, 17, 36-37, 65, 109, 113, 115
optical dissociation, 23
oxidative stress, 41
oxybenzone, 17, 18, 33, 34, 49, 65, 66, 72, 113, 124
oxygen radical scavengers, 111

P

o-PABA, 65
[6-4] photoproducts, 51, 52
p-aminobenzoate, 15, 38, 65, 106, 159
p-Aminobenzoic acid, 14-15, 33-34, 38, 50-51
p21(Cip1/Waf1), 66, 180-181
p53 induction, 179-180
p539, 179
Padimate-O, 18, 33-34, 36, 50-55, 182
Parsol-1789, 33
penetration depth in skin, 3
permeability of human skin, 59
peroxidation, 26, 30, 182
phosphorescence, 22, 28
photo-Ames test, 182
photoaging, 13, 69-71, 81, 85, 88-90, 93-95, 127
photocarcinogenesis, 79, 145, 157-158, 165, 168-169, 172
photodermatoses, 118
photoexcitation, 23
photofragmentation, 23, 28-33, 41-42
photosensitized pyrimidine dimerization, 38
photosensitizer, 25, 35-37, 40, 64
physical blockers, 14, 26-28, 41
physiological assays, 181
piperidine, 52
polymorphic light eruption, 118
polypeptide, 86
polysaccharides, 87
promoter, 85, 89-95, 103, 177
prostaglandins, 105, 158
proteins, 1, 37, 86-87, 182
PUVA, 94-95, 101
pyrimidine photohydrates, 37

Q

quencher of free radicals, 55

R

radical anion, 26, 30-31

S

S. cerevisiae, 51
Salmonella, 51
scavenger, 41-42, 54, 111
sensitized reactions, 183
short-term efficacy, 178
singlet molecular oxygen, 30, 35-36, 111
singlet-singlet energy transfer, 25
skin carcinogenesis, 99-100
skin iron content, 182
solar simulating radiation, 69, 71, 79, 83, 91
solar simulator, 32, 36, 63, 65-66, 71-72, 108, 130, 132-137, 140-143, 146-150
solar urticaria, 118
solar UV spectrum, 159
solar-simulated UV radiation, 112, 159
solvated electron, 30, 33
SPF 7, 18, 71, 73-81
squamous cell carcinoma, 8, 48, 100, 102
SSUV (solar simulated UV), 113, 159, 165–170, 166–170
stirred absorbing solution, 2
stratospheric ozone, 13
Sun Protection Factor (SPF), 63, 69, 91, 99, 159, 185
sun-damaged, 88-90
sunburn, 14, 48-49, 63, 65, 69, 71, 73-74, 79-80, 99, 106-107, 110, 116-118, 127, 130, 132, 148, 150, 158, 178-181, 184-185
sunlamps, 65, 107-108, 110, 112, 114-115, 133-149, 185
sunlight, 1, 6, 13, 28, 31-32, 35, 42, 47-49, 50, 53, 56-57, 63, 66, 69-70, 80, 86, 100, 102, 106, 112, 127-128, 130, 134-136, 141-146, 149-150, 158-159, 165, 177, 180-184
sunscreen, 63-66, 69, 70-81, 91-95, 99, 106-108, 110-114, 116-119, 127-128, 130, 132-133, 136, 138-139, 142-143, 145-146, 148-149, 157-169, 171-172
  commercial, 32, 43, 57-58, 66, 111, 113, 128, 132, 159–164
sunscreen protected skin (SPS), 179
sunscreen testing, 130, 132, 150
sunscreen-DNA photoadducts, 34
superbasal layer, 180
superoxide, 26, 30-31
superoxide radical anion, 26, 30-31
suppressor T lymphocytes, 102, 104
suppressor T-cells, 128
systemic CH, 128, 132, 134, 137, 149, 160, 162-163, 167

T

T4 endonuclease V, 116, 117
thymine, 32, 34, 37-40, 50-52, 70
thymine dimer, 37-39
thymine peroxyl radical, 32
thymine radical, 32

titanium dioxide, 18, 27-28, 40, 55-59, 106, 109-111, 159, 161
transcription factors, 179
transmission of skin, 2, 7
transplant recipients, 100
triplet states, 25, 49
triplet-triplet energy transfer, 25
tumor necrosis factor-α, 105
tumor progression, 177
tumor susceptibility, 112, 162
tumor-associated-antigens, 145
tumor-immune suppression, 145
tumor-immunity, 129, 134
Type I collagen, 85-86
Type III collagen, 86-87

U

U.S. Food and Drug Administration, 14, 128
ultraviolet, 11, 13-14, 23, 37, 42, 53, 63, 65, 69, 85, 90-91, 99, 115, 127, 179, 182-183
ultraviolet radiation, 109
Urocanic Acid, 1, 4, 105, 116-117, 123, 125, 158, 163-164, 166-167, 169-172
  cis-isomer, 35
  photoisomerization, 24, 28, 32-35, 41-42, 116, 163-164, 166-167
UV dose-response curves, 145
UV power spectra, 128, 134-135, 141-143
UV radiation, 13, 20, 28-29, 69, 70, 72, 99-106, 112-113, 116, 157-159, 162, 164-165, 181
UV source, 128-137, 140-144, 146-149, 150
UV-inducible protein, 178
UV-regressor tumors, 143
UVA, 4, 8, 13, 19, 26-29, 41, 48-49, 56-57, 63-65, 69, 70-73, 76, 79, 81, 91-92, 94-95, 105-106, 110-111, 113, 118, 122, 134-135, 140-143, 149, 158-159, 165-167, 173, 179, 180, 182-184
UVA I, 80
UVA II, 79, 81
UVB, 2, 3, 8, 13, 18, 20, 27-29, 35, 37, 39, 41, 48, 50, 52, 56, 57, 63-66, 69-72, 79-81, 91-95, 100, 101, 103-107, 110-113, 117, 118, 133-147, 149, 158-168, 171-172, 179-180, 182-184
UVB-Absorber, 111, 165, 171
  2-EHMC (2-ethylhexy-p-methoxycinnamate), 34-35, 40, 65, 106-107, 110-113, 118, 159, 161, 163-165, 167-172
  O-PABA (octyl-N-dimethyl-p-aminobenzoate), 65
  V, 40
UVC, 11, 13, 112, 143, 179

V

Verhoeff-van Gieson, 88
Versican, 88-89
vibrational relaxation, 21, 23, 28

X

xenon arc lamps, 32
xeroderma pigmentosum, 6, 100-101, 145

Z

zinc oxide, 27-28, 55-59, 106, 110-111